FRONTIERS OF SCIENCE

MARINE SCIENCES

Notable Research and Discoveries

KYLE KIRKLAND, PH.D.

Facts On File
An imprint of Infobase Publishing

MARINE SCIENCES: Notable Research and Discoveries

Facts On File, Inc.
An imprint of Infobase Publishing
132 West 31st Street
New York NY 10001

Library of Congress Cataloging-in-Publication Data

Kirkland, Kyle.
  Marine sciences : notable research and discoveries / Kyle Kirkland.
    p. cm.—(Frontiers of science)
  Includes bibliographical references and index.
  ISBN 978-0-8160-7443-3
  1. Marine sciences. 2. Oceanography. 3. Marine biology. I. Title.
  GC16.K57 2010
551.46—dc22   2009026065

Facts On File books are available at special discounts when purchased in bulk quantities for businesses, associations, institutions, or sales promotions. Please call our Special Sales Department in New York at (212) 967-8800 or (800) 322-8755.

You can find Facts On File on the World Wide Web at http://www.factsonfile.com

Text design by Kerry Casey
Illustrations by Melissa Ericksen
Photo research by Tobi Zausner, Ph.D.
Composition by Mary Susan Ryan-Flynn
Cover printed by Bang Printing, Inc., Brainerd, Minn.
Book printed and bound by Bang Printing, Inc., Brainerd, Minn.
Date printed: May 2010
Printed in the United States of America

10 9 8 7 6 5 4 3 2 1

This book is printed on acid-free paper.

CONTENTS

PREFACE

Discovering what lies behind a hill or beyond a neighborhood can be as simple as taking a short walk. But curiosity and the urge to make new discoveries usually require people to undertake journeys much more adventuresome than a short walk, and scientists often study realms far removed from everyday observation—sometimes even beyond the present means of travel or vision. Polish astronomer Nicolaus Copernicus's (1473–1543) heliocentric (Sun-centered) model of the solar system, published in 1543, ushered in the modern age of astronomy more than 400 years before the first rocket escaped Earth's gravity. Scientists today probe the tiny domain of atoms, pilot submersibles into marine trenches far beneath the waves, and analyze processes occurring deep within stars.

Many of the newest areas of scientific research involve objects or places that are not easily accessible, if at all. These objects may be trillions of miles away, such as the newly discovered planetary systems, or they may be as close as inside a person's head; the brain, a delicate organ encased and protected by the skull, has frustrated many of the best efforts of biologists until recently. The subject of interest may not be at a vast distance or concealed by a protective covering, but instead it may be removed in terms of time. For example, people need to learn about the evolution of Earth's weather and climate in order to understand the changes taking place today, yet no one can revisit the past.

Frontiers of Science is an eight-volume set that explores topics at the forefront of research in the following sciences:

- biological sciences
- chemistry
- computer science

- Earth science
- marine science
- physics
- space and astronomy
- weather and climate

The set focuses on the methods and imagination of people who are pushing the boundaries of science by investigating subjects that are not readily observable or are otherwise cloaked in mystery. Each volume includes six topics, one per chapter, and each chapter has the same format and structure. The chapter provides a chronology of the topic and establishes its scientific and social relevance, discusses the critical questions and the research techniques designed to answer these questions, describes what scientists have learned and may learn in the future, highlights the technological applications of this knowledge, and makes recommendations for further reading. The topics cover a broad spectrum of the science, from issues that are making headlines to ones that are not as yet well known. Each chapter can be read independently; some overlap among chapters of the same volume is unavoidable, so a small amount of repetition is necessary for each chapter to stand alone. But the repetition is minimal, and cross-references are used as appropriate.

Scientific inquiry demands a number of skills. The National Committee on Science Education Standards and Assessment and the National Research Council, in addition to other organizations such as the National Science Teachers Association, have stressed the training and development of these skills. Science students must learn how to raise important questions, design the tools or experiments necessary to answer these questions, apply models in explaining the results and revise the model as needed, be alert to alternative explanations, and construct and analyze arguments for and against competing models.

Progress in science often involves deciding which competing theory, model, or viewpoint provides the best explanation. For example, a major issue in biology for many decades was determining if the brain functions as a whole (the holistic model) or if parts of the brain carry out specialized functions (functional localization). Recent developments in brain imaging resolved part of this issue in favor of functional localization by showing that specific regions of the brain are more active during certain tasks. At the same time, however, these experiments have raised other questions that future research must answer.

The logic and precision of science are elegant, but applying scientific skills can be daunting at first. The goals of the Frontiers of Science set are to explain how scientists tackle difficult research issues and to describe recent advances made in these fields. Understanding the science behind the advances is critical because sometimes new knowledge and theories seem unbelievable until the underlying methods become clear. Consider the following examples. Some scientists have claimed that the last few years are the warmest in the past 500 or even 1,000 years, but reliable temperature records date only from about 1850. Geologists talk of volcano hot spots and plumes of abnormally hot rock rising through deep channels, although no one has drilled more than a few miles below the surface. Teams of neuroscientists—scientists who study the brain—display images of the activity of the brain as a person dreams, yet the subject's skull has not been breached. Scientists often debate the validity of new experiments and theories, and a proper evaluation requires an understanding of the reasoning and technology that support or refute the arguments.

Curiosity about how scientists came to know what they do—and why they are convinced that their beliefs are true—has always motivated me to study not just the facts and theories but also the reasons why these are true (or at least believed). I could never accept unsupported statements or confine my attention to one scientific discipline. When I was young, I learned many things from my father, a physicist who specialized in engineering mechanics, and my mother, a mathematician and computer systems analyst. And from an archaeologist who lived down the street, I learned one of the reasons why people believe Earth has evolved and changed—he took me to a field where we found marine fossils such as shark's teeth, which backed his claim that this area had once been under water! After studying electronics while I was in the air force, I attended college, switching my major a number of times until becoming captivated with a subject that was itself a melding of two disciplines—biological psychology. I went on to earn a doctorate in neuroscience, studying under physicists, computer scientists, chemists, anatomists, geneticists, physiologists, and mathematicians. My broad interests and background have served me well as a science writer, giving me the confidence, or perhaps I should say chutzpah, to write a set of books on such a vast array of topics.

Seekers of knowledge satisfy their curiosity about how the world and its organisms work, but the applications of science are not limited to intellectual achievement. The topics in Frontiers of Science affect society on a multitude of levels. Civilization has always faced an uphill battle to procure scarce resources, solve technical problems, and maintain order. In modern times, one of the most important resources is energy, and the physics of fusion potentially offers a nearly boundless supply. Technology makes life easier and solves many of today's problems, and nanotechnology may extend the range of devices into extremely small sizes. Protecting one's personal information in transactions conducted via the Internet is a crucial application of computer science.

But the scope of science today is so vast that no set of eight volumes can hope to cover all of the frontiers. The chapters in Frontiers of Science span a broad range of each science but could not possibly be exhaustive. Selectivity was painful (and editorially enforced) but necessary, and in my opinion, the choices are diverse and reflect current trends. The same is true for the subjects within each chapter—a lot of fascinating research did not get mentioned, not because it is unimportant, but because there was no room to do it justice.

Extending the limits of knowledge relies on basic science skills as well as ingenuity in asking and answering the right questions. The 48 topics discussed in these books are not straightforward laboratory exercises but complex, gritty research problems at the frontiers of science. Exploring uncharted territory presents exceptional challenges but also offers equally impressive rewards, whether the motivation is to solve a practical problem or to gain a better understanding of human nature. If this set encourages some of its readers to plunge into a scientific frontier and conquer a few of its unknowns, the books will be worth all the effort required to produce them.

ACKNOWLEDGMENTS

Thanks go to Frank K. Darmstadt, executive editor at Facts On File, and the rest of the staff for all their hard work, which I admit I sometimes made a little bit harder. Thanks also to Tobi Zausner for researching and locating so many great photographs. I also appreciate the time and effort of a large number of researchers who were kind enough to pass along a research paper or help me track down some information.

INTRODUCTION

Spanish explorer Vasco Núñez de Balboa (1475–1519) became the first known European to see the Pacific Ocean after leading his expedition across the Isthmus of Panama in 1513. But it was Portuguese explorer Ferdinand Magellan (1480–1521) who gave this ocean its name, *Mar Pacífico*, meaning "peaceful sea." A relieved Magellan described it as "peaceful" after rounding the stormy southern tip of South America in 1520 and finding calm weather as he sailed northward, into the Pacific Ocean.

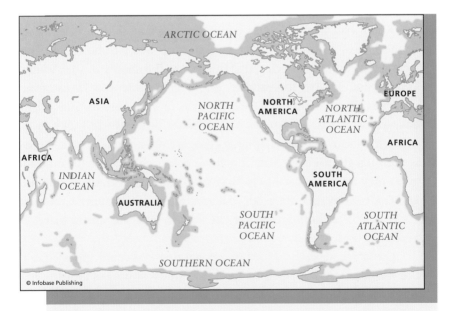

This map shows the four major oceans as well as the Southern Ocean—shallow depths near the continents are shown in darker shade.

But people who live around the Pacific Ocean know that this body of water is not always so calm. Covering about a third of the planet, the Pacific Ocean has held many surprises for the explorers and scientists who came after Magellan. It continues to do so for researchers today.

The other oceans are just as interesting, and all are interconnected—a world ocean, as it is sometimes called, divided into separate seas for geographical or historical reasons. Magellan's crew could attest to the ability to sail all the way around the world, after an arduous voyage begun in 1519 and completed in 1522. (Magellan died in the Philippines in 1521, and only one ship out of the original five made it home.) The figure shows a flattened (and therefore somewhat distorted) view of the globe. There are five named oceans, if one counts the Southern or Antarctic Ocean. They cover about 70 percent of Earth's surface.

Marine Sciences, one volume of the multivolume Frontiers of Science set, is about explorers and scientists who expand the frontiers of marine science—and often find things they do not expect. The study of the ocean is known as oceanography or marine science. Included is the study of the geology of the seafloor, the chemical and physical properties of the water, and the life that teems in and around it. This book discusses six topics that touch on many of the frontiers of marine science.

About half of the world's population lives in coastal areas. Many of the largest cities in the United States are close to the ocean; New York City and Los Angeles are situated on the shores of the Atlantic and Pacific Oceans, respectively, and Philadelphia and Houston are quite near the ocean as well. (Chicago sits on the shore of Lake Michigan, one of the largest bodies of freshwater in the world.) One of the most important reasons that large cities are often found near the coast is transportation—a deep harbor allows ocean-going ships to come and go as they travel to distant lands, carrying goods for trade. Economics and trade motivated Magellan, Italian seafarer Christopher Columbus (1451–1506), and other explorers to venture onto the seas in search of profitable trade routes.

Explorers needed safe ships, navigational aids, and knowledge of the weather, which helped to spark the study of oceans and the marine environment. But just as important, and perhaps more so, was the urge to make new scientific discoveries and find out what lies hidden beneath the waves. Curiosity is a powerful incentive for explorer and scientist alike. Economics and the search for profits may have funded

Scientists and artists created this image of Earth from a collection of satellite images and data. Earth's oceans are prominent features of the planet. *(NASA images by Reto Stöckli, based on data from NASA and NOAA)*

the voyages of Magellan and other explorers, but the drive to find out what lies beyond the ordinary realm must have surely played a role in the willingness of voyagers to take such grave risks.

Marine scientists have made progress, but there are still many frontiers awaiting exploration. Each chapter of this book explores one of these frontiers. Reports published in journals, presented at conferences, and described in news releases illustrate research problems of interest in marine science, and how scientists are tackling them. This book discusses a selection of these reports—unfortunately there is room for only a fraction of them—that offers students and other readers insight into the methods and applications of oceanography.

Students need to keep up with the latest developments in these quickly moving fields of research, but they have difficulty finding a source that explains the basic concepts while discussing the background and context essential for the "big picture." This book describes the evolution of each of the six main topics it covers, and explains the problems that researchers are currently investigating as well as the methods they are developing to solve them.

Chapter 1 describes how researchers have taxed their endurance and devised sophisticated instruments to help them reach the ocean floor. In 1960, researchers Donald Walsh (1931–   ) and Jacques Piccard (1922–2008) descended in a cramped *submersible* to the deepest known part of any ocean—35,800 feet (10,912 m) in a section of the Pacific Ocean called the Mariana Trench. Marine scientists have learned much about the ocean floor and its geology and natural resources by diving in manned submersibles or by operating remotely controlled vehicles. Other opportunities are becoming available for researchers to "see" what lies beneath the waves with the use of satellite imagery. Yet much of the seabed remains to be explored.

One of the most impressive discoveries concerning the ocean floor is a globe-circling volcanic feature known as the *mid-ocean ridge.* Chapter 2 explains how scientists found this extraordinary structure, and how their explanation of its origins was a tremendous advance in geology as well as oceanography. The study of this mid-ocean ridge has also furthered other scientific disciplines in a surprising manner—certain properties of nearby volcanic rocks offer a historical record of Earth's *magnetic field,* and biologists have discovered, much to their astonishment, thriving communities of unusual organisms that take advantage of the heat and minerals that spew from the volcanic vents.

In addition to volcanoes, waves, and variations that influence the world's weather, the ocean also harbors an astounding variety of life. The astonishment of finding organisms around the hot volcanic vents has been repeated in several other fascinating and unexpected discoveries. Chapter 3 describes the discovery of giant squid and the deep dives of the whales that hunt them, and the eerie light-emitting organisms that inhabit the cold, dark, eat-or-be-eaten depths of the ocean.

The topic of chapter 4 involves one of the reasons why the Pacific Ocean is not always so peaceful. Tsunami, a Japanese term that means harbor wave, is a series of waves that result when a large body of water is suddenly displaced. Such displacements frequently happen when undersea earthquakes move a large volume of rock in the ocean floor. When a tsunami reaches shore, the waves attain enormous heights, sweeping onto the shore and causing catastrophic flooding. The Pacific Ocean is home to many tsunami-generating earthquakes and is the focus of much research. But other areas are also susceptible, as tragically demonstrated in 2004 when an Indian Ocean tsunami claimed 250,000 victims.

Along with tsunamis, the Pacific Ocean also features a phenomenon that is less violent but even wider in scope. El Niño is a warming of waters off the South American coast that occurs irregularly every few years. (The name is Spanish for little boy, a reference to the birth of Christ, since the phenomenon tends to begin around Christmas.) As discussed in chapter 5, researchers have discovered an association between El Niño and wind patterns across the tropical Pacific Ocean, as well as links to weather and climate in regions as far away as the United States.

Most life in the sea congregates in the upper part, in the shallow areas that receive plenty of sunlight. Plants use light from the Sun to make carbohydrates, and these plants in turn provide food for other organisms. Microscopic plants called *algae* are critical to life in the ocean because they form the base of food chains, but algae populations can rise to dangerously high numbers. Chapter 6 discusses blooms, which are sudden explosions of population growth that in certain cases can release toxic substances and poison the environment. People have been inadvertently contributing to this situation because nutrients such as excess fertilizer are washing down rivers and fueling some of these blooms.

Algal blooms, tsunamis, and El Niño-related weather disruptions are important challenges to people who live on the world's coasts, as well as to marine science. And all of the topics in *Marine Sciences* have vital applications to economics and the environment. The topics also display the tremendous breadth and scope of the frontiers of marine science.

THE OCEAN DEPTHS: EXPLORING THE SEABED

On August 23, 1960, United States Navy Lieutenant Donald Walsh (1931–) and Swiss scientist Jacques Piccard (1922–2008) made the deepest dive in history. In the bathyscaphe *Trieste*—named for the Italian city where Jacques's father Auguste Piccard (1884–1962) designed the vessel—the two men reached a depth of 35,800 feet (10,912 m)—6.8 miles (10.9 km)—in a region of the Pacific Ocean called the Mariana Trench. (The term *bathyscaphe* comes from the Greek words *bathys,* deep, and *skaphē,* small boat.) The Mariana Trench contains the deepest known part of the oceans, called the Challenger Deep, first thoroughly surveyed in 1951 by HMS *Challenger.*

The dive was punctuated at one point by an ominously loud cracking noise—a Plexiglas viewing window suddenly fractured due to the tremendous pressure at this depth, which exceeds the pressure at sea level by a factor greater than 1,000. But the cracked window did not yield, and the divers returned to the surface after about nine hours. Most of this time was spent on the way down and the way back up; Piccard and Walsh were on the bottom only 20 minutes, shivering in the 45°F (7.2°C) temperature of the vessel's cabin. (Outside, the water temperature at that depth was about 37°F [2.8°C].) In the 1961 book *Seven Miles Down,* Piccard and R. S. Dietz wrote, "Indifferent to the nearly 200,000 tons of pressure clamped on her metal sphere, the *Trieste* balanced herself delicately on the few pounds of guide rope that lay on the bottom, making token claim, in the name of

1

science and humanity, to the ultimate depths in all our oceans—the Challenger Deep."

Perhaps surprisingly, the Mariana Trench is not in the middle of the Pacific Ocean. Instead, it is located near the Mariana Islands (from which it takes its name) in the northwestern region of the ocean. The ocean floor, or seabed, is not featureless or bowl-shaped, but contains deep trenches, tall "mountains" called *seamounts* (which, if they reach the surface, are known as islands), and the majority of the world's volcanic activity. Exploring the seabed is a fascinating topic for many reasons, including its status as one of the last frontiers of Earth.

This chapter describes what researchers have learned about the ocean floor. These studies have yielded much information about the history of Earth—especially valuable are studies of the mid-ocean ridge (an area of much geological activity where new crust appears)—and recent tsunamis have focused attention on the seabed and undersea earthquakes. Natural resources such as oil have also been found underneath the ocean. The seabed is home to many exotic creatures and

View from a submersible of the seafloor at a depth of about 1,310 feet (400 m) off the coast of North Carolina *(Dr. Steve Ross, UNC-W. NOAA Office of Ocean Exploration)*

extremophiles (organisms that live in extreme environments) as well as the remains of numerous kinds of sunken ships, including cruise ships such as the RMS *Titanic,* galleons carrying gold and other treasure, and nuclear submarines.

Much remains to be discovered. Only a few percent of the ocean floor has been mapped in detail—scientists know more about the surface features of the Moon, Mars, and Venus than the ocean. This lack of knowledge creates hazards and accidents, such as on January 8, 2005, when the submarine USS *San Francisco* slammed at full speed of about 35 miles (56 km) per hour into an uncharted seamount, killing one crewman and nearly sinking the vessel. But the relatively unknown frontier of the ocean floor also creates many opportunities for researchers to learn more about Earth and its magnificent oceans.

INTRODUCTION

The earliest recorded attempt at deep-sea *bathymetry*—measuring the depth—was made in ancient Greece. In 85 B.C.E., Posidonius (ca. 135–50 B.C.E.), a Greek philosopher, sailed into the Mediterranean Sea and dropped a rock attached to a long rope. After paying out about 6,000 feet (1,830 km) of rope, the rock stopped falling, indicating it had reached bottom. This technique, called sounding, was the technique that sailors used for the next 2,000 years to determine depth.

Sounding with a weighted rope or wire is simple and effective in shallow depths. All the sailor has to do is to extend the rope slowly and wait for the weight to touch bottom, at which time the rope can be marked at the surface and then hauled up. The depth of the sea at that point is the length of the rope from the weighted end to the mark. Depths were often described in units known as *fathoms,* which was the distance of the sailor's outstretched arms and was counted as the sailor hauled up the rope. (The term *fathom* comes from an Old English word, *fœthm,* meaning the length of outstretched arms.) A fathom has since been standardized to equal six feet (1.83 m).

Problems occurred when sailors attempted to measure deeper depths. Long ropes have a lot of mass, and underwater currents twirl and push the rope and weight. A sailor cannot discern the moment of touch in deep water, and might continue to pay out line, which settles on the bottom. The additional slack causes an overestimation of the depth. Results from deep-sea soundings before the 19th century varied wildly, and scientists had little confidence in ocean bathymetry until late in that century.

In 1872, British scientist Sir William Thomson (1824–1907) (later to become Lord Kelvin) developed a sounding machine that spooled out thin wire. This machine sensed the resistance of the dropping weight, improving the accuracy of the soundings. Often the spool would stop well before a person would have, but skeptics could confirm that the weight had reached the bottom after the machine retrieved the line—if the sinker had touched bottom, it would be covered in the gray muddy ooze of the ocean floor. Various other sounding machines were developed around this time, and in 1872 the HMS *Challenger* set sail for a 3.5-year voyage, the first major expedition whose primary function was to study the ocean. (Although it shares the name, this ship was not the same vessel that surveyed the Mariana Trench in the 1950s.) Scientists on the HMS *Challenger* retrieved a small amount of seabed sediment with Baillie sounding machines. Named after Navigating Lieutenant C. W. Baillie, the sinker had a tube that captured a small amount of mud as it struck the ocean floor, allowing the researchers to study samples in the ship's laboratories.

Making a sounding with rope or wire and a sinker, especially of extremely deep water, is a slow, laborious process. But there is a faster way. Water, as the Greek philosopher Aristotle (384–322 B.C.E.) noticed more than two millennia ago, conducts sound. The conduction of sound is much more rapid in water than air—the speed of sound in seawater is about 3,350 miles (5,400 km) per hour, slightly less than five times the speed of sound in air under normal conditions. Early oceanographers, such as American Naval expert and researcher Matthew Fontaine Maury (1806–73), suggested in the middle of the 19th century the use of sound for bathymetry. The idea was to bounce sound waves off the seabed and determine from the reflections the distance and shape of the ocean floor. But a lack of equipment to detect faint sound waves hampered the development of sound-based bathymetry.

War and military technology provided the impetus for improving the equipment, as is the case for many scientific techniques. Detecting submarines became essential during World War I (1914–18). Researchers such as French physicist Paul Langevin (1872–1946) and others engineered transmitters and sensitive receivers using quartz crystals.

In 1922, using a new sonic depth-finder, the USS *Ohio* cruised from New York to Chesapeake Bay, recording depths up to about 1,700 fathoms. These devices calculated the depth and topography of the seabed by using the time required for sound to reflect off the bottom and arrive at the receiver. Later, in 1925, the German vessel *Meteor* made the first extensive voyage using echo sounding and discovered a long ridge running along the middle of the Atlantic Ocean—a mid-ocean ridge, the subject of chapter 2.

Sonar, which stands for sound navigation and ranging, is the general term for the use of sound for underwater navigation and detection. (The term *sonar* is similar to radar—which stands for radio detection and ranging, a technique to locate and identify objects by their reflection of electromagnetic radiation called radio waves or microwaves.) Perfected and named during World War II (1939–45), sonar has been commonly used in bathymetry since the 1950s. Animals such as dolphins and bats also use sound waves to find prey or navigate in the dark, a process known as echolocation.

Using these techniques, plus others described in the following two sections, marine scientists have learned that the average depth of the oceans is about 12,234 feet (3,730 m), although estimates vary because only a small percentage of the ocean floor has been mapped in detail. The Pacific Ocean is, on average, slightly deeper than the others, at about 13,000 feet (3,960 m). Compare these figures to the average height of the continents, which is 2,755 feet (840 m). The average depth of the ocean is about four and a half times greater than the average rise of the continents above sea level.

As shown in the figure, the ocean floor is not a smooth, bowl-shaped gap between the continents. Extending a varying distance beyond the continents is a shallow stretch known as the *continental shelf,* which consists mostly of rocks similar to those of the continents. On average, the continental shelf extends about 40 miles (64 km) into the ocean, although in some locations, such as parts of the northern coast of North America, the shelf reaches up to nearly 1,000 miles (1,600 km). The gently descending continental shelf breaks at the continental slope, which falls at an average angle of about 4 degrees, but can reach a 25-degree angle. In deeper waters, the ocean floor levels out into an *abyssal plain,* connected to the continental slope by a short stretch known as the continental rise, which is often filled with rocks and debris that have

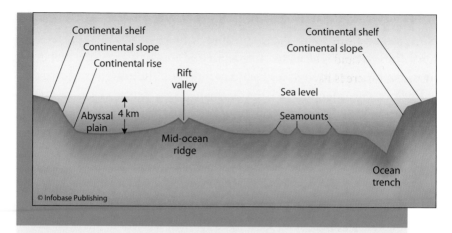

A cross section through the ocean displays the varied geological features of the seabed.

tumbled down the slope. (The term *abyss,* which generally refers to any vast space or depth, comes from a Greek word, *abyssos,* meaning bottomless.) Features of the ocean floor include ridges and seamounts that rise toward the surface, and trenches that plunge to greater depths.

DEEP-SEA SUBMERSIBLES

Sounding and sonar are tools to extend human knowledge of the ocean by remote sensing—detection and analysis at a distance. To learn even more, researchers would like to visit the seabed and get a close-up view. But the only way to reach the forbidding environment of the deep ocean is in a deep-sea submersible such as a bathyscaphe, a vehicle designed to withstand tremendous pressure.

Pressure due to the atmosphere is a familiar part of life on Earth's surface. The air of the atmosphere consists of gases—mostly nitrogen and oxygen along with a little water vapor, carbon dioxide, and traces of other substances—that are trapped by the planet's gravity. Air does not weigh much, but it can exert substantial effects as it resists the movement of cars or airplanes, which reduces the fuel efficiency of these vehicles and, at high speeds, raises the temperature of the vehicular surface due to frictional heating. Even when an object is at rest, air molecules continually bombard it, creating a pressure known as atmospheric pressure. At the surface of the planet, or in other words at sea level (the ocean surface when calm), the pressure can be measured as a weight or force acting on a specific area, and is equal to 14.7 pounds/inch^2 (1 kg/cm^2, or 101,325 P). The atmosphere is like an "ocean" of air, and any object at the "bottom," such as a person standing on Earth's surface at sea level, feels the weight of a column of air above, pressing down on him or her.

Water is much denser than air and exerts even more force on a given area. As a diver descends 33 feet (10 m) beneath the ocean's surface, the water above exerts a pressure equal to about one atmosphere, which means the diver's body experiences a pressure equal to two atmospheres (the water plus the air above it). At a depth of 100 feet (30 m), a diver experiences a pressure four times as strong as he or she would at sea level. This may not seem like much, but consider that the additional pressure of the water feels like 44.1 pounds (20 kg) pressing against each square inch (6.25 cm^2) of the diver's body. That is a lot of weight to bear! Farther down, the massive pressure would crush a diver.

Submarines are also limited in how far they can dive, and even the sturdiest submarines in the United States Navy are unable to withstand the pressure at depths greater than about 2,400 feet (730 m). The pressure at this depth would probably crush the hull. (The term *probably* is needed here—submarine crews are understandably reluctant to confirm this estimate, which is based on the manufacturer's standards.)

For deep dives, a specially constructed vehicle is essential, and divers must bring their own lights, for little sunlight can penetrate the great depths. Some of the earliest deep-sea vessels were hollow spheres made of steel, known as bathyspheres. In 1934, oceanographer William Beebe (1877–1962) and engineer Otis Barton (1899–1992) descended 3,028 feet (923 m) in the Atlantic Ocean near the island of Bermuda. (The shape of a sphere is excellent for this purpose because it distributes the pressure uniformly.) *Trieste,* which launched in 1953 and carried Piccard and Walsh on their record dive in 1960, consisted of flotation tanks and a spherical chamber underneath where there was just enough space for two people. The figure shows a drawing of this vessel. The tanks were filled with gasoline, which is lighter than water and therefore buoyant, along with water that could be pumped in or out for ballast. Operated initially by the French navy, the vessel was purchased by the U.S. Navy in 1958. Navy officials retired the *Trieste* in 1966, but the craft is on display at the Washington Navy Yard in Washington, D.C.

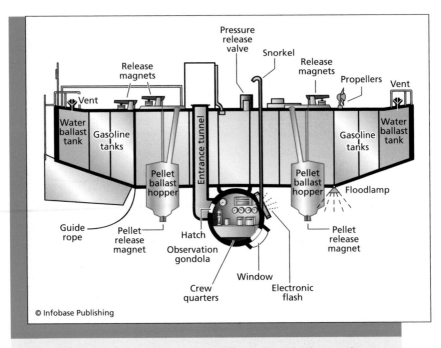

The diagram of the *Trieste* shows flotation tanks and other equipment in the body of the vessel, and space for two cramped people in the sphere underneath the middle of the craft.

The United States Navy has operated many other deep sea submersibles, a class of vehicles the Navy refers to as DSV (Deep Submergence Vehicle). *Trieste* was designated DSV-0, and *Trieste II*, built in 1964, is DSV-1. DSV-2 is *Alvin,* the construction of which was funded by the Office of Naval Research (ONR), which is responsible for the science and technology programs of the United States Navy. Commissioned in 1964 and soon to retire, *Alvin* is a highly maneuverable craft that the Woods Hole Oceanographic Institution (WHOI) in Massachusetts currently owns and operates. (The name *Alvin* honors Allyn Vine [1914–94], an engineer at WHOI who designed the vessel.) This vehicle has been used to investigate many underwater phenomena, including the *hydrothermal vents* described in chapter 2, deep-sea life described in chapter 3, and the RMS *Titanic* wreck.

The Navy's oceanographic research helps its mission to defend the United States, as well as advancing science. The following sidebar offers some more information on the Office of Naval Research.

Alvin [R. Catanach/OAR/National Undersea Research Program [NURP]; Woods Hole]

Office of Naval Research

The role of science in World War II was immense. Anti-submarine finders, mine detectors, bombs that skip across water, and many other devices, including the war-ending atomic bomb, made vital contributions to the war effort. Recognizing this contribution, the United States Navy was one of the first government organizations to fund university researchers, who previously had to rely on uncertain funding from college departments to equip their laboratories and pay the salaries of technicians. In May 1945, near the end of the war, the Secretary of the Navy created an office for Navy research, and Congress later passed legislation to establish the Office of Naval Research (ONR), signed into law, Public Law 588, by then-president Harry Truman on August 1, 1946.

ONR's mission, according to Public Law 588, was "to plan, foster, and encourage scientific research in recognition of its paramount importance as related to the maintenance of future naval power and the preservation of national se-

A Japanese submersible, *Shinkai 6500,* is the champion diver among submersibles active at the present time. This vehicle, operated by the Japan Agency for Marine-Earth Science and Technology, accommodates three passengers and can attain a depth of 21,320 feet (6,500 m), while *Alvin* can only reach 14,760 feet (4,500 m). (The maximum depth in meters explains the number in *Shinkai 6500,* and the word *shinkai* is Japanese for "deep sea.") It is being used to study deep-sea life as well as the sudden movements on the ocean floor responsible for tsunamis. As yet, no one has a modern, maneuverable submersible that can revisit the deepest part of the ocean—the Challenger Deep—although Hawkes Ocean Technologies, a company in California, is testing a design, called Deep Flight, that might work.

Mimicking the *National Aeronautics and Space Administration* (NASA), which often sends unmanned probes to visit planets and

curity." In October of its first year, ONR invested about 22 million dollars in 400 different projects. Since then, ONR has sponsored projects that have led to numerous advances, such as the work of Charles H. Townes (1915–), who developed the maser—microwave amplification by stimulated emission of radiation—the forerunner of the laser. ONR has also funded research on hurricanes, Arctic environments, deep-sea submersibles such as *Alvin*, radio telescopes, robots, and much more.

In addition to funding research at institutes and universities, ONR operates its own laboratory, the Naval Research Laboratory, located in Washington, D.C. This laboratory predates ONR, since it was established on July 2, 1923, and later transferred to ONR's management. Projects at the Naval Research Laboratory have been crucial in the development of radar, satellites (including GPS, the global positioning system), advanced materials for fighting fires, and many other technological applications. Research at this laboratory, as well as the ONR-sponsored projects at various institutes and universities across the country, have clear benefits extending well beyond military operations.

moons located vast distances away from Earth, some marine scientists use remotely operated vehicles to explore the deep sea. The advantage of vessels operated from the safety of a surface ship is that the submersible need not maintain life support systems or the rigorous safety standards necessary for human passengers. Remotely operated vehicles can be equipped with cameras for taking pictures and robotic arms to retrieve samples. Although there are no passengers to feel the excitement and awe of the dive, the vehicles usually perform admirably. But accidents sometimes happen. On May 29, 2003, the Japanese remotely operated vessel *Kaiko* was exploring a trough in the Pacific Ocean at a depth of 15,330 feet (4,675 m), when the craft experienced a power failure, and a cable attached to a rover assembly broke. The rover became lost, and subsequent efforts to find it were unsuccessful.

Another important tool to study the ocean floor is also an unmanned vehicle, but its realm of operation is entirely different. Researchers are now using satellites orbiting high above the planet to map the ocean floor and study its terrain.

MAPPING THE OCEAN FLOOR FROM SPACE

The reason that scientists presently have better maps of the surface of the Moon and some of the distant planets than the ocean floor is the high-quality photographic capability of space-faring probes that make the long journey. Satellites orbiting Earth have also mapped its surface features in great detail, but the vast amount of water in the oceans hides the seabed from view. But the newest technology and highest precision instruments have begun to surmount this obstacle, giving oceanographers a sweeping view of the depths.

How can satellites see through so much water? The answer is that they do not see through the water, they measure small variations in the level of the ocean's surface. These variations reflect the topography of the ocean floor.

Water normally takes the shape of its container and spreads itself smoothly over its features, maintaining an even surface except for periodic disturbances such as waves. For the most part, this is true of water in the ocean, which maintains its level around the globe. Variations in the ocean surface due to wind, which whips up waves, and tides, which are due to the gravitational pull of the Moon (and to a lesser extent, the Sun), are easily observed.

But other variations in the surface of the ocean exist, although they are less pronounced than wind or tidal waves. Consider a large seamount on the ocean floor. This large mass alters the gravitation field in its vicinity by a small amount. Although the gravitational effect is slight, the fluidity of water means that it can respond to the minute variations. The additional gravity attracts water, which bulges at the surface of the ocean over the seamount. The bulge is difficult to detect because it covers a broad area, has a gentle slope, and at its peak may only be one foot (0.3 m) or so above the height of the surrounding sea.

The United States Navy satellite *Geostat*, launched in 1985, and the European Space Agency satellite *ERS-1*, launched in 1991, housed radar altimeters, which are instruments capable of making highly precise measurements of altitude. These satellites emitted electromagnetic radiation and detected the reflections from the planet's surface. Because

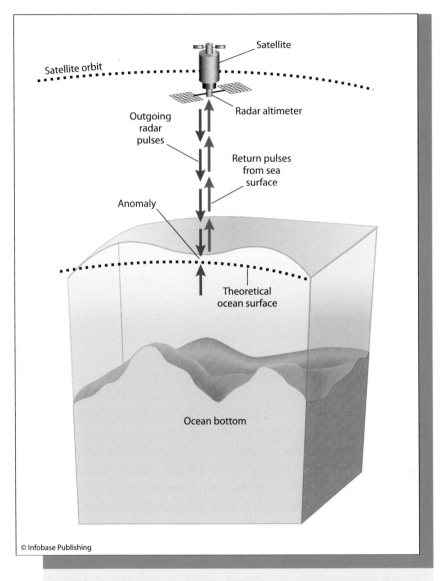

© Infobase Publishing

Satellites use radar altimeters to measure the water level with great precision, and these measurements can detect deviations or anomalies from the expected level.

the speed of the radiation is known (in air, approximately 186,000 miles [300,000 km] per second), the satellites could determine the distance traveled by using the time required for the round trip. Other equipment, such as networks of lasers, determined the exact altitude of the satellite. From these two measurements, oceanographers could map the level of the surface of the ocean to within about 1 inch (2.5 cm). The figure illustrates the process.

To deal with the variations due to waves and tides, researchers needed to make a lot of measurements of the same area. In some of those measurements, a wave crest increases the height of a certain spot, while in other measurements, a wave trough decreases the height. By averaging a large number of measurements, these random variations cancel out—the crests and troughs of the waves tend to add to zero—and what is left is the actual level of the ocean.

Data from the satellites were not widely available until recently (some of the data were classified), but upon their release, Walter Smith, a researcher at the *National Oceanic and Atmospheric Administration* (NOAA), and David Sandwell of Scripps Institution of Oceanography in California began to create maps. Bumps in these maps indicate features of the ocean floor, as confirmed by comparing the satellite maps to those obtained from submersibles and sonar. The researchers published their paper, "Global Sea Floor Topography from Satellite Altimetry and Ship Depth Soundings," in a 1997 issue of *Science*.

Satellite bathymetry gives oceanographers a "bird's-eye view" of the ocean floor, just as satellite imagery has aided geologists and astronomers studying the land surface of Earth and other solar system bodies. The technique also greatly speeds up the generation of ocean floor maps, making a vast improvement on the much slower methods that rely on sonar and submersibles—as Smith and Sandwell relate in their paper, "Conventional sea floor mapping is a tedious process." Improvements in satellite bathymetry will give researchers an even better view of the ocean floor, showing more details. For example, *Jason-2*, a satellite named after the mythological Greek mariner who sailed the *Argo*, launched on June 20, 2008. The satellite, a combined project of NASA, NOAA, European Organisation for the Exploitation of Meteorological Satellites, and Centre National d'Études Spatiales (National Center of Space Research, a French space agency), carries advanced altimetry equipment and will collect a tremendous amount of data in the coming years.

GEOLOGICAL PROCESSES THAT SHAPE THE OCEAN FLOOR

The ridges, trenches, and undersea mountains of the ocean floor, as revealed during submersible dives, echo sounding and sonar, and satellite altimetry, led marine scientists to wonder what produced these features. The answer involves the important geological topic of plate tectonics.

In 1912, German scientist Alfred Wegener (1880–1930) proposed the startling idea that continents drift. Some continents seemed to fit together like pieces in a jigsaw puzzle; for instance, the west coast of South America is a good match for the east coast of Africa. Wegener theorized that continents drift over time, and that once, long ago, the continents of the world were collected into one gigantic land mass, which subsequently broke apart. Skeptics could not imagine how something as large as a continent could move, so the continental drift theory was not taken seriously at the time.

But as geologists of the 20th century studied Earth's structure, they discovered the planet consists of different layers. These layers have different compositions and temperatures, which scientists discovered in the early 1900s when they began to measure and analyze *seismic waves* that are generated during earthquakes. (The composition, temperature, and physical state of rocks affect the conduction and speed of these waves, which seismologists can measure with instruments called *seismometers.*) Earth has a solid metal inner core, a liquid metal outer core, a rocky mantle, and a thin layer of crust on top. The temperature of the core is not precisely known, but it is hot, around 14,400°F (8,000°C). Upper layers get progressively cooler. (The heat is leftover from Earth's fiery creation, plus a contribution from ongoing *radioactive* decay, which emits a lot of energy.)

The physical properties of Earth's interior show that the rocks at the surface, consisting of the crust and the thin, uppermost portion of the mantle, are rigid. This rigid layer is known as the lithosphere, and extends from Earth's surface down to about 62 miles (100 km). (The term *lithosphere* derives from *lithos*, a Greek word for stone.) Beneath it, the mantle is hot and partially molten, forming a slushy or plastic layer known as the asthenosphere that extends from the lithosphere down to about 435 miles (700 km). (Asthenosphere gets its name from *asthenēs*, a Greek word meaning weak.)

Although the lithosphere is rigid, it does not consist of a giant spherical shell surrounding the globe, but is instead broken into about 12 major plates, along with a few dozen smaller ones. These plates are known as *tectonic plates*. (The term *tectonic* derives from a Greek word, *tekton,* meaning "of a builder.") Tectonic plates are not fixed in place because they slowly glide over the partially molten asthenosphere. The speed varies from plate to plate and over time, but it is generally in the range of 1–6 inches (2.5–15 cm) per year.

Plate motion alters the face of the planet, albeit extremely slowly, because the continents get rearranged over time. Wegener was right— millions of years ago, the continents were joined together. The motion of plates also has more immediate effects. Plate speed and direction of motion varies, which means that some plates are diverging (separating), some plates are converging (colliding), and some plates grind along another plate's border. Stress builds up along plate boundaries, resulting in sudden movements that create earthquakes. Fissures known as faults, such as the San Andreas Fault in California, arise along plate boundaries and are the sites of many of the world's earthquakes. Cracks along plate boundaries also allow molten rock known as *magma* to seep up from below, forming volcanoes. The vast majority of Earth's volcanic activity and earthquakes occurs along plate boundaries.

In the ocean, the separation of plates creates the undersea feature known as the mid-ocean ridge. For example, in the southern portion of the Atlantic Ocean, the South American plate and African plate are pulling away from each other at a rate of about 1.4 inches (3.5 cm) each year. Through the crack flows magma, which solidifies and forms a new section of ocean floor. In addition to the new crust along the ocean floor, this volcanic activity produces a ridge running along the boundary. This volcanic ridge exists throughout the world's oceans and is known as the mid-ocean ridge—although it is not necessarily in the middle of the ocean—forming kind of a seam that runs around the globe. (The portion of the ridge that exists in the Atlantic Ocean is approximately in the middle, and is known as the Mid-Atlantic Ridge.) The new crust spreads as the plates slowly move apart, as first noted by American geologist Harry Hess (1906–69) in the 1960s. About three-quarters of the world's volcanic activity occurs along the mid-ocean ridge. Mid-ocean ridges and the unique environment they create is the subject of chapter 2 of this book.

ABE pencil beam sonar image of Explorer Ridge
4 meter grid-cell size
3 times vertically exaggerated
Looking north up the rift valley

Magic
Mountain
Area

Sonar image of an ocean ridge—the Southern Explorer Ridge, located in the Pacific Ocean off the coast of British Columbia, Canada, including Magic Mountain, an area of hydrothermal activity (see chapter 2) *(NOAA)*

When tectonic plates collide instead of separate, one plate usually dives under another. This often occurs when ocean plates, which consist of dense crust, dives or subducts beneath the lighter continental crust. (The rocks of the continents are light enough to rise above sea level, forming the world's landmasses.) As one plate dives beneath the other, a deep trench is created, as shown in the figure on page 6. The Mariana Trench, for example, is found where the Pacific Plate dives beneath the Philippine Plate. Heat in Earth's interior melts the rocks of the diving plate, which can rise to the surface and create volcanoes. Movements along plate boundaries also create earthquakes under the sea as well as on land, with an additional danger in the ocean of triggering a dangerous series of waves known as a tsunami.

The constant motion of plates, the bubbling up of new crust in divergence zones, and the melting of the rocks of subducting plates produces a cycle known as the Wilson cycle, named for Canadian geologist John Tuzo Wilson (1908–93). New crust forms at mid-ocean ridges, becomes part of an ocean plate, then moves slowly, eventually experiencing another plate boundary, where it may get pushed underneath and melt—recycling the material—or it may get pushed upward, forming part of a continent. Mountain ranges such as the Himalaya in Asia, including Mount Everest, contain marine fossils in rocks that were once part of the seabed! The age of ocean floor crust varies, with areas close to the ridges being more recent. Because of the cycle, the oldest ocean crust is only about 180 million years old, quite recent in terms of Earth's geology (Earth formed about 4.6 billion years ago).

Studying plate motion and the associated geological processes continues to be an active area of research in geology as well as marine science. One ambitious project led by John Delaney of the University of Washington aims to construct a network of sensors on the ocean floor, connected by fiber optic cables, over a broad area of the Juan de Fuca Plate. This relatively small plate, located off the coast of Washington, sits between the Pacific Plate and the North American Plate, and plays a vital role in the active geological region of the west coast and Cascade Mountains. The project, initially called NEPTUNE (named after the Roman god of the sea), is now part of the Ocean Observatories Initiative of the National Science Foundation, a United States government agency that funds basic research. A network of sensors will allow widespread monitoring of plate movement, seismic activity, organisms, and chemical reactions associated with the ocean floor. Delaney and his colleagues expect construction to begin by 2010 and to be completed five or so years later. The data gathered in this project will help scientists understand the mechanics of undersea plate movements as well as their various chemical, geological, and biological impacts on the ocean floor environment.

RESOURCES LOCATED BENEATH THE OCEAN FLOOR

The study of geological processes such as plate tectonics helps scientists to understand ocean floor topography, but the substances composing the floor—and the area directly underneath—are also interesting. Particles, including bodies of dead plants and animals along with other debris, fall to the ocean floor, forming a layer of oozy mud known as *sediment*. The ocean's water exerts a lot of pressure on these sediments.

Over time, the sediments solidify, producing sedimentary rocks that make their way across oceans and into continents during the Wilson cycle.

To study sediments, marine scientists analyze the composition and biological activity of samples retrieved from submersibles. Sediment properties vary; sediments known as lithogenous derive mostly from rock (*lithos*) and contain much quartz or clay (the term *lithogenous* derives from *lithos* plus *genos,* a Greek word meaning birth or origin), while hydrogenous sediments contain material that precipitates from the water due to chemical reactions. (*Hydōr* is a Greek word for water, which is the basis for the term *hydrogenous* as well as the term *hydrology,* the study of water.) Lithogenous sediments can often be found on the continental shelf, slope, and rise, as well as in some regions of the deep ocean. Certain hydrogenous sediments are often found on the abyssal plain as well as shallow areas of the continental shelf.

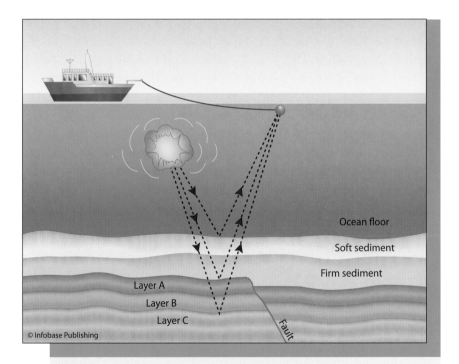

© Infobase Publishing

Researchers study the layers beneath the seabed by creating underwater waves and listening for the echoes.

How do researchers study what lies beneath the ooze? One method is to use seismic waves, the same technique that geologists use to study Earth's interior. To study shallow depths, geologists and marine scientists sometimes create their own waves, as shown in the figure. Scientists studying land surfaces will often make seismic waves with large trucks that shake the ground with an oscillating pad or metal disk, while seismic investigations under the ocean tend to make a seismic disturbance by releasing compressed air. A wave striking the boundary between any two materials will be partially reflected and partially transmitted, as can be seen with light waves striking glass—some light is transmitted, allowing a viewer to see what is on the other side, but some light is reflected, providing a (normally) faint image of the viewer. The same is true for seismic waves encountering boundaries between rocks of different densities, or between rock layers and a pool of oil.

With seismic techniques, geologists can identify promising areas under the ocean to drill for oil. Oil and natural gas deposits exist in many underground locations all over the world, although drilling in the ocean floor presents difficulties because of the water, and there are also concerns about oil escaping into the marine environment. The first offshore oil well, which Henry L. Williams and his associates constructed in 1897, was located on a pier on the coast of California, near Santa Barbara. (Edwin Drake built the first successful oil well on land at Titusville, Pennsylvania, in 1859.) Offshore oil production usually takes place on specially constructed platforms floating in the ocean; drills on these platforms can operate in water at depths of about 2 miles (3.2 km). In the United States, most offshore oil drilling occurs in the Gulf of Mexico, though some platforms are still located off the California coast. Ocean drilling accounts for about a quarter of the oil production in the United States.

Oil and other petroleum products are not the only resources found on or under the seabed. In 2006, for instance, scientists from the Woods Hole Oceanographic Institution and Nautilus Minerals, Inc., a company headquartered in Toronto, Canada, explored the Pacific Ocean floor off the coast of Papua New Guinea. The team of researchers used remotely operated vehicles to search for sites rich in minerals such as copper and gold. They found some potentially rich sites, and in 2007, Nautilus Minerals began ordering drilling equipment. The company's plan is to begin operating the world's first seabed copper and gold mine in 2010, at a site in 4,920 feet (1,500 m) of water located about 31 miles (50 km) north of

Rabaul, Papua New Guinea. A seafloor-mining tool will do the drilling, and pump the ore up to a support vessel on the surface.

The ocean floor is also an important resource for scientists who want to learn how Earth and its oceans evolved. Buried within the layers of sediments is a lot of history; the chemistry and composition of a sediment layer reflects the temperature and conditions of the ocean in which it was deposited. The sediments accumulate slowly, and the deeper sediments are the oldest, having been deposited in earlier times. By drilling into the seabed and removing a vertical sample—a core—scientists can analyze the layers and gain valuable clues about the environment at the time the sediments were deposited.

For example, the Arctic Coring Expedition, sponsored by an international research program known as the Integrated Ocean Drilling Program, obtained a 1,400-foot (428-m) core of sediment in 2004 from the central Arctic Ocean. By studying the oxygen compounds in the core, along with other sediment data, researchers propose that the Arctic region made a transformation from a land-locked lake to a fully oxygenated, well-circulated ("ventilated") ocean about 17.5 million years ago, in a geological epoch called the Miocene. The circulation, which brought oxygen and nutrients into the waters, may have been due to a newly formed connection with the Atlantic Ocean. Prior to this time, the Arctic sea would have been isolated. The researchers, led by Martin Jakobsson of Stockholm University in Sweden, published their findings, "The Early Miocene Onset of a Ventilated Circulation Regime in the Arctic Ocean," in a 2007 issue of *Nature*.

Reconstructing the history of the oceans and their chemical activity is important in understanding how Earth and its marine environment evolved. This information is vital if scientists are to understand how climates have changed over time, which is essential for a better understanding of the global climate changes, such as global warming, that are currently taking place.

SUNKEN VESSELS—WOOD, IRON, TREASURE CHESTS, AND NUCLEAR REACTORS

Ocean floors accumulate sediment gradually, except every once in a while a large chunk of something falls from the surface or upper layers of the sea. Dead whales and other carcasses descend to the ocean bottom, providing an important source of food for deep sea creatures. In recent times, human technology has added another class of object, at least when something goes wrong—sunken vessels. The International Registry of Sunken Ships lists more than 112,000 ships as of June 2009. The United Nations Educational, Scientific, and Cultural Organization estimates that a total of more than three million shipwrecks lie on the seafloor.

What happens to these sunken vessels? The tremendous pressure crushes the ships that sink into the great depths. Vessels sinking in relatively shallow waters may retain their shape, but all objects in the ocean are subject to various chemical reactions. Marine environments include not only water but also substances that are dissolved in it.

The composition of seawater varies across oceans and depths, but some amount of dissolved materials are always present. (This is also true in most freshwater, although there are generally much higher concentrations of salt, for instance, in seawater.) The *salinity* of water measures the amount of dissolved inorganic matter (that is, matter not associated with living organisms), and is equal to the ratio of the mass of the dissolved substances to the mass of the water. Most of this material comes from salts—compounds whose components are joined by chemical bonds called ionic bonds, formed when atoms share electrons. In water, salts dissolve into ions. One of the most common salts—found generally at the dinner table—is sodium chloride. These compounds have collected in oceans due to absorption from the seabed, runoff from land surfaces, and continual evaporation of water from the oceans (which leaves behind the dissolved material). Typical seawater has a salinity of 3.5 percent, meaning that there are 3.5 units of dissolved matter for every 100 units of water. More than a quarter of the dissolved material comes from sodium and chloride ions, but seawater contains at least trace amounts of about 80 chemical elements.

The earliest ships were made of wood, which was replaced by steel beginning in the 19th century. Wood decays quickly in seawater because of chemical reactions with the dissolved material as well as the activity of marine organisms. Ancient shipwrecks are generally already gone—in many cases, all that is found is some of the cargo—unless the ship was buried in sediment, protecting it from attack. For instance, the *Mary Rose,* flagship of Henry VIII that launched in 1511, was found in about 45 feet (14 m) of water off the coast of England, much of the wreck covered in clay. Engineers raised the preserved portion of the vessel in 1982, and displayed the *Mary Rose* in Portsmouth, England—one of the oldest wooden warships on display in the world.

Marine environments are also not kind to steel. Steel is an alloy of iron with a small amount of carbon and, in some cases, other elements. Metals containing iron are vulnerable to corrosion. Iron readily combines with oxygen—much of Earth's iron is found in such compounds—and in the presence of oxygen and water, iron undergoes a chemical reaction that produces an iron oxide commonly known as rust. The ions in seawater accelerate the process, so iron and steel corrodes more quickly in or near the ocean. Although metal hulls, cannons, cannon balls, anchors, and other thick objects can survive for long periods, eventually the corrosion weakens the structure, and the metal and oxides get swept away by the flow of water currents.

One of the most famous ships to sink was the RMS *Titanic.* This luxury cruise ship began her maiden voyage from Southampton, England, on April 10, 1912, with 2,240 passengers and crew, bound for New

The bow of the sunken RMS *Titanic (AP Photo/Nauticus)*

York. But on the night of April 14, as immortalized in Walter Lord's 1955 book (and 1958 film), *A Night to Remember,* the RMS *Titanic* hit an iceberg and sank in the North Atlantic Ocean. Ship designers erroneously considered the RMS *Titanic* unsinkable, so the crew was not prepared for disaster; the vessel carried an insufficient number of lifeboats, and in the ensuing chaos, some of the lifeboats were launched with plenty of empty seats, leaving even more passengers and crew without the means of rescue. The water was so cold that an immersed person suffered hypothermia—a drastic cooling of body temperature—and died shortly thereafter. About 1,500 people perished in the tragedy.

The sunken ship lay undisturbed on the bottom of the ocean until 1985, when an expedition led by Jean-Louis Michel and Robert Ballard discovered the wreck. This site, some 375 miles (600 km) from Newfoundland, was about 14 miles (22 km) from the ship's last known position. The depth is about 12,500 feet (3,800 m). The ship is mostly intact, except that a section of the rear of the ship lies separated from the front, about 2,000 feet (610 m) away.

How long will the sunken RMS *Titanic* last in its watery grave? Although corrosion of the ship's steel hull would ordinarily be a vital factor, the requirement for oxygen and water becomes important at certain ocean depths. There is abundant water, of course, but at the depths of the Atlantic Ocean where the RMS *Titanic* rests, oxygen is in short supply because of the lack of sunlight for marine plants to perform *photosynthesis* (one of the products of which is oxygen). Some oxygen drifts down from the upper ocean, but without plentiful oxygen in the vicinity of the sunken vessel, corrosion is limited.

Yet the hulking wreck is deteriorating, mainly because of the activity of microorganisms that are able to metabolize—digest—iron and other components of the ship's hull. The result is similar to rusting. Estimates on how long the wreck of the RMS *Titanic* will remain range from about 100 to 400 years. According to Lori Johnston, a Canadian researcher quoted by the British Broadcasting Corporation (BBC) in an interview dated October 18, 2005, "We predict in between 80 and 100 years you will probably still see the U-shaped hull but all the decks will have collapsed in."

Exploring sunken ships lets divers learn about history and revisit the scene of frightening or momentous disasters, such the sinking of a large cruise ship. Such exploration can also be profitable.

After Columbus discovered America in 1492, Spain and other countries sent ships to the new world and found significant quantities of precious metals, among other resources, which the colonists took—on some occasions "looted" would be a better word—and shipped home. Spanish ships sailed for home weighed down with fabulous treasure, but crossing the ocean in those wooden ships was extremely risky, and some of them failed to make it. For example, the *Nuestra Señora de Atocha* (Our Lady of Atocha) encountered a hurricane and sank on September 6, 1622, near Key West, Florida. This ship was laden with much silver, gold, and other goods. Explorer and treasure hunter Mel Fisher and his team found what little remained of the wreck in 1985 (the same year RMS *Titanic* was discovered). But the treasure was immense, valued at about 450 million dollars (gold is a relatively inert metal and usually survives immersion fairly well).

The rights to a sunken vessel can be a thorny issue, testing the intricacies of courts and international laws, depending on who found the ship, who owned the ship when it sank, where it sank, and when. But often the issue boils down to finders' keepers. The argument over *Nuestra Señora de Atocha* began even before the wreck and its treasure were found; Fisher, who spent years searching for the sunken vessel, and the governments of the United States and Florida engaged in a bitter dispute over ownership. The United States Supreme Court decided in 1982 that Fisher would own the treasure if he found it, which he did three years later.

Sometimes environmental issues become the paramount issue in a sunken vessel. Such is the case when a nuclear-powered submarine sinks. Strong, corrosion-resistant enclosures protect the nuclear reactor that supplies power to these vessels, but some people worry that the integrity of these shields may be compromised when the vessel sinks or is exposed to years of marine environments. If so, some radioactive material may leak, exposing the surrounding area of the ocean to dangerous levels of radiation. Eight nuclear-powered submarines—two American and six Russian—have sunk as of June 2009, so the question is relevant.

In one case, the sunken vessel was raised. A consortium of Dutch companies and the Russian Navy raised the Russian submarine *K-141 Kursk*, which sank on August 12, 2000. For wrecks that remain on the floor, testing of the water around the sunken submarines has shown no sign of radioactive materials. For instance, the latest tragedy was the sinking of the Russian submarine *K-159* on August 30, 2003 (as the vessel was being towed toward a harbor). A few months later, Bellona, a nonprofit environmental foundation based in Oslo, Norway, monitored the area around the sunken submarine and detected no significant leaks. But as a safety precaution, the ocean floor and the water in the vicinity of these wrecks will probably continue to be monitored.

LIVING ON THE OCEAN FLOOR

Life on the ocean floor is hard enough without having to contend with disasters such as sunken submarines. Thanks to the extensive pressure of the water and a scarcity of sunlight, life would seem to have a precarious foothold at great depths. But certain kinds of living organisms have been able to adapt to these harsh conditions.

Many species of microorganisms such as bacteria thrive under inhospitable circumstances, and some of these species are able to make a meal from material that is entirely indigestible to other creatures. The bacteria that are presently devouring RMS *Titanic,* for instance, "eat" metal, and other bacteria devour elements that are equally unpalatable to humans. In the vicinity of the mid-ocean ridge, hot water and minerals escape from vents on the seabed, providing nutrients to certain microorganisms. Larger organisms have evolved to dine on the small ones, creating a fascinating ecosystem full of strange and unusual animals.

Marine scientists have been astounded at the extent of life's spread into the ocean. In 2003, researchers led by James P. Cowen at the University of Hawaii reported finding microorganisms thriving in the crust about 984 feet (300 m) beneath the bottom of the ocean! The crust, located on the flanks of the Juan de Fuca Ridge, is fractured, with hot water seeping through it. It seems that living organisms on Earth inhabit any place in which they can possibly survive. The researchers published their report, "Fluids from Aging Ocean Crust That Support Microbial Life," in a 2003 issue of *Science.*

CONCLUSION

Mapping the bottom of the ocean has gone from soundings with a weighted rope, a technique as old as ancient Greece, to the use of sound waves and then satellite radar. Marine researchers have found the ocean floor a rich, dynamic place, full of life as well as geological activity, including the majority of the world's volcanism. From the continental shelf to the deepest trenches, people have explored the terrain, learned something about the ocean, and developed the technology to find and extract oil, minerals, and other resources.

But the seafloor is still a vast frontier of marine science. The movement of plates, the mechanisms of undersea earthquakes and volcanoes, and the evolution of the oceans are still not fully understood. And the chemistry of the seabed, and its variety of life-forms, are just now coming to light.

Further study of the ocean floor may include long-term visits. Space researchers have ventured off the planet and have established an orbiting habitat, the International Space Station, to maintain a permanent (more or less) presence in space. Oceanographers may one day do the same at the great depths of the ocean.

Manned underwater stations have already been tried and developed, though they have a spotty history. In 1964, the United States Navy initiated the SEALAB project, which consisted of a series of underwater environments. SEALAB I, lowered 193 feet (59 m) to the bottom in July 1964, near the coast of Bermuda, housed a team of four "aquanauts." They lived there for 11 days, until a tropical storm forced them to evacuate the chamber. In the next phase of the experiment, engineers built and placed SEALAB II in 205 feet (62 m) of water off the California coast in 1965. The structure, 57 feet (17.4 m) by 12 feet (3.7 m), provided living quarters for 10-person teams, including Scott Carpenter, who had also gone into space as an astronaut in the Mercury project. Helping the SEALAB II aquanauts was a trained dolphin named Tuffy that performed errands and transported supplies from the surface.

One of the most important features of the SEALAB program was to test the endurance of humans living on the ocean floor. The pressure even at these relatively shallow depths is immense. Problems would have occurred if the interior of the lab was held at atmospheric pressure, since an aquanaut entering the chamber would have experienced a sudden change in pressure, which causes decompression sickness and severe pain, occasionally even death. (Decompression sickness, also known as the "bends," is due to dissolved gases suddenly bubbling out of the tissues because of the rapid change in pressure.)

To avoid decompression problems, the air pressure of the lab was the same as that of surrounding water. Raising the air pressure meant pumping more air into the chamber, so that there was a much greater density of gas molecules than on the surface of the earth, which resulted in a much higher "atmospheric" pressure in the lab. Because exposing humans to gases such as nitrogen or oxygen at high pressure for long periods is not generally healthy, helium was used. The helium-rich air was safe to breathe, though it made the aquanauts' voices sound squeaky.

The first two labs were a success, but SEALAB III, a more ambitious phase of the project, failed. In 1969, SEALAB III descended to 610 feet (186 m) off San Clemente Island, California. A leak quickly developed. Divers were unable to repair the chamber, and aquanaut Berry Cannon died in the attempt. The Navy scrapped the project shortly thereafter.

Only a few underwater habitats are operating today. Aquarius, a National Oceanic and Atmospheric Administration laboratory, is located at a depth of 60 feet (18.3 m) in the Florida Keys National Marine Sanctuary, and Jules' Undersea Lodge, named for writer Jules Verne (1828–1905), is a habitat in 30 feet (9 m) of water near Key Largo, Florida, that bills itself as the "first and only underwater hotel."

Deep sea habitats may not be feasible now or in the near future because of the extreme pressure, but there is 10,000,000 miles^2 (26,000,000 km^2) of continental shelf available. This is an area larger than the continent of North America. As the human population on Earth continues to expand, the need for additional room and resources becomes an increasingly serious problem. Oceanographers and adventurers have begun to explore the vast territory of the ocean floor, and the advancement of marine science provides a greater understanding of this region of the planet, as well as opportunities to use it wisely.

CHRONOLOGY

85 B.C.E.   Ancient Greek philosopher Posidonius (ca. 135–50 B.C.E.) measures the bottom of the Mediterranean Sea.

1872–76 C.E.   HMS *Challenger,* the first major expedition to study the ocean, sails around the world.

1897   Henry L. Williams and his associates build the first offshore oil well, located on a pier on the California coast.

1922   USS *Ohio,* using one of the first depth-finders based on sound waves, cruises from New York to Chesapeake Bay, recording depths up to about 1,700 fathoms.

1925   German vessel *Meteor* makes the first extensive voyage using echo sounding, and discovers the mid-ocean ridge.

1951   HMS *Challenger* (often signified *Challenger* II, distinguishing it from the earlier ship), explores the deepest known part of the ocean, at the Mariana Trench.

1960   United States Navy Lieutenant Donald Walsh (1931–) and Swiss scientist Jacques Piccard (1922–2008) descend in the bathyscaphe *Trieste* 35,800 feet (10,912 m) into the ocean at the Mariana Trench.

1960s   American geologist Harry Hess (1906–69) recognizes that seafloor spreading is occurring at the mid-ocean ridges.

1964   The United States Navy conducts underwater habitat experiments in SEALAB I.

1965   The United States Navy continues its underwater habitat research with SEALAB II.

1969   The United States Navy attempts to extend the underwater habitat experiment to a depth of 610 feet (186 m), but the operation is unsuccessful.

1985   Jean-Louis Michel, Robert Ballard, and their crew find the wreck of the RMS *Titanic.*

Mel Fisher and his team find the treasure of *Nuestra Señora de Atocha,* a Spanish vessel that sank off the coast of Florida in 1622.

1997	Walter Smith and David Sandwell begin using satellite data to make maps of the ocean floor.
2007	Nautilus Minerals, Inc., launches plans to build and operate the first seabed copper and gold mine.
2008	The satellite *Jason-2,* the combined effort of the National Aeronautics and Space Administration and the National Oceanic and Atmospheric Administration in the United States, and the European Organisation for the Exploitation of Meteorological Satellites and Centre National d'Études Spatiales (National Center of Space Research, a French space agency), makes a successful launch. This advanced satellite will collect a tremendous amount of data concerning the oceans.

FURTHER RESOURCES
Print and Internet

Ballard, Robert D. *The Discovery of the Titanic.* New York: Time Warner, 1987. One of the most famous sea disasters, RMS *Titanic,* an "unsinkable" cruise ship on her maiden voyage, collided with an iceberg on April 14, 1912, and sank. In 1985, a team led by Jean-Louis Michel and Robert Ballard found the wreck lying at a depth of about 12,500 feet (3,800 m). Ballard tells the story of this extraordinary undersea expedition in this book.

British Broadcasting Corporation. "Ocean Reclaiming Titanic Liner." October 18, 2005. Available online. URL: http://news.bbc.co.uk/1/hi/northern_ireland/4352568.stm. Accessed June 9, 2009. This news article describes the deterioration of the wreck of the RMS *Titanic.*

Broad, William J. *The Universe Below: Discovering the Secrets of the Deep Sea.* New York: Touchstone, 1998. Broad, a science writer, takes readers on a tour of the ocean depths, describing dives, divers, submersibles, and shipwrecks. A chronology of deep sea exploration is included.

Cowen, James P., Stephen J. Giovannoni, Fabien Kenig, H. Paul Johnson, David Butterfield, Michael S. Rappé, et al. "Fluids from Aging Ocean Crust That Support Microbial Life." *Science* 299 (January 3, 2003): 120–123. The researchers report the discovery of microorganisms thriving in the crust about 984 feet (300 m) beneath the bottom of the ocean.

Jakobsson, Martin, Jan Backman, Bert Rudels, Jonas Nycander, Martin Frank, Larry Mayer, et al. "The Early Miocene Onset of a Ventilated Circulation Regime in the Arctic Ocean." *Nature* 447 (June 21, 2007): 986–990. Data presented in this paper indicates that the Arctic region made a transformation from a land-locked lake to a fully oxygenated, well-circulated ocean in a geological epoch called the Miocene.

Matsen, Bradford. *Descent: The Heroic Discovery of the Abyss.* New York: Vintage, 2006. This is the story of William Beebe and Otis Barton as they risked their lives in their bathysphere to be the first to see and experience the great depths of the ocean.

Piccard, Jacques, and Robert S. Dietz. *Seven Miles Down.* New York: Putnam, 1961. The authors describe the descent to the deepest part of the ocean.

Public Broadcasting Service. "Lost Liners." Available online. URL: http://www.pbs.org/lostliners/. Accessed June 9, 2009. This Web resource, based on the book *Lost Liners* by Robert D. Ballard and Rick Archbold, tells the story of five famous shipwrecks—*Titanic, Lusitania, Empress of Ireland, Britannic,* and *Andrea Doria*—and contains information on the development of the great ocean liners.

Sandwell, David T., and Walter H. F. Smith. "Exploring the Ocean Basins with Satellite Altimeter Data." Available online. URL: http://www.ngdc.noaa.gov/mgg/bathymetry/predicted/explore.HTML. Accessed June 9, 2009. Sandwell and Smith developed a method of mapping the ocean floor using satellite radar data. On this well-illustrated page, the authors discuss how it works, and what they have learned about the bottom of the ocean.

Scripps Institution of Oceanography. "Satellite Geodesy." Available online. URL: http://topex.ucsd.edu/marine_topo/. Accessed on June 9, 2009. This Web resource features an interactive map of the ocean floor, along with pages devoted to the tools of satellite imagery of the oceans, including radar altimetry and synthetic aperture radar.

Smith, Walter H. F., and David T. Sandwell. "Global Sea Floor Topography from Satellite Altimetry and Ship Depth Soundings." *Science* 277 (September 26, 1997): 1,956–1,962. The researchers use satellite data to generate seafloor maps.

University of Delaware College of Marine Studies. "A Deep-Sea Odyssey." Available online. URL: http://www.ocean.udel.edu/extreme2001/home/. Accessed June 9, 2009. Climb aboard the submersible *Alvin* and take a virtual tour of an expedition to hot water vents on the ocean floor.

———. "To the Depths in *Trieste*." Available online. URL: http://www.ocean.udel.edu/extreme2004/geology/trieste.html. Accessed June 9, 2009. This Web page contains a short but fascinating article describing the adventure of Jacques Piccard and Donald Walsh as they made their record descent in the bathyscaphe *Trieste* in 1960.

University of Rhode Island. "Discovery of Sound in the Sea." Available online. URL: http://www.dosits.org/. Accessed June 9, 2009. Sound waves are critical for submarine navigators, oceanographers studying the ocean floor, and dolphins hunting their prey. The Office of Marine Programs at the University of Rhode Island maintains this Web resource, which is full of articles, photographs, and other resources describing the study of sound in the ocean.

Web Sites

Integrated Ocean Drilling Program. Available online. URL: http://www.iodp.org/. Accessed June 9, 2009. The Integrated Ocean Drilling Program is an international effort to explore the structure and history of Earth's evolution, as detailed in ocean floor sediments. The program's Web site contains information on expeditions, reports, plans, and schedules for future projects.

International Registry of Sunken Ships. Available online. URL: http://www.shipwreckregistry.com/. Accessed June 9, 2009. Reports of sunken ships are available from this Web site, usually for a small fee.

Office of Naval Research. Available online. URL: http://www.onr.navy.mil/. Accessed June 9, 2009. The Office of Naval Research conducts a great deal of marine research as part of their ongoing efforts to support Navy operations. Their Web site offers information on past and present projects, science, and technology.

2

Mid-ocean Ridge— The Largest Single Volcanic Feature on the Planet

At an ocean depth of 8,000 feet (2,440 m), the temperature is generally about 39°F (4°C), and the pressure is about 240 times as strong as it is at sea level. In 1977, scientists and explorers including Robert Ballard—the discoverer of the wreck of the RMS *Titanic*—made a dive to this depth in an area of the Pacific Ocean known as the East Pacific Rise, near the Galápagos Islands. Descending in the *Alvin,* a deep-sea submersible described in chapter 1, the researchers became the first to see a hydrothermal vent, from which issued a steady stream of hot water. (The term *hydrothermal* comes from the Greek words *hydōr,* meaning water, and *thermē,* heat.) Although the hydrothermal vent was not a big surprise, what the researchers found in the vicinity was amazing—an environment teeming with life, including giant tube worms.

Finding a hot water "spring" or vent was a fascinating discovery in itself, although the researchers were aware of underwater volcanic activity in this area, which made hydrothermal vents a possibility. But the extent to which living organisms had colonized this environment was not expected. In these frigid depths, where sunlight fails to penetrate, the hot water and its mineral content form an undersea "oasis" that supports a variety of organisms.

This chapter explores these special environments and their inhabitants. The Pacific Rise is one section of a vast ocean floor geological formation known as the mid-ocean ridge. This feature rises above the surrounding seafloor and contains a central rift or valley. The ridges zigzag across the oceans, forming a jagged "seam," and are the sites of earthquakes and about 75 percent of the world's volcanic activity. Researchers have learned much about the mid-ocean ridge since its discovery in the 20th century, but the effects and impact of this undersea volcanic region are not yet clear. Its influence on marine biology, especially in and around the hydrothermal vents, is enormous, making this an important frontier of marine science.

INTRODUCTION

Dates obtained by comparing radioactive atoms in rocks indicate that Earth formed about 4.5 billion years ago. Although Earth was too hot in its earliest stages to support liquid water on the surface, oceans formed as the planet cooled and water collected from the water vapor emissions of volcanic activity, possibly along with contributions from comet impacts. Earth has probably had seas for about 4 billion years of its existence.

But such an old age for the world's oceans leads to an interesting puzzle. Sediment consisting of small particles or biological materials continually falls to the ocean floor, accumulating layers on the seabed that gradually get compressed into sedimentary rock due to the pressure of the overlying water. Four billion years is an awfully long time. With all that sediment falling and accumulating on the bottom, the oceans would probably have filled up by now.

One possible solution to this puzzle is that the seafloor is not as old as the oceans themselves. What this means is that there is some mechanism to replace the ocean floor over a period of time. In this scenario, the sediments and sedimentary rock of the ocean floor are continually recycled; some portion of the floor moves above sea level or else is destroyed, and some portion is generated to replace it. There is some evidence for such turnover in the seafloor, since sedimentary rock and marine fossils have been found on land, even as high as the Himalaya Mountains! But prior to the middle of the 20th century, scientists did not know what sort of mechanism could be responsible.

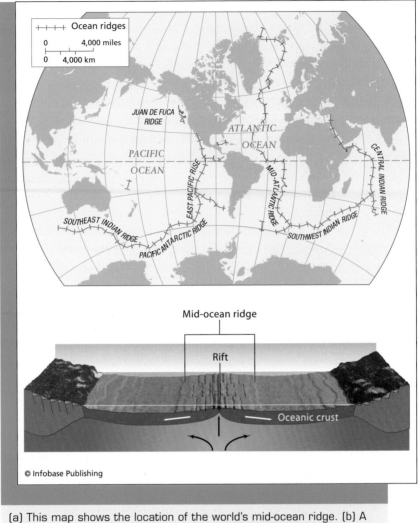

(a) This map shows the location of the world's mid-ocean ridge. (b) A cross section through the ridge shows the elevated portion and the central valley, along with hot, molten rock rising underneath.

A clue came in 1912, when German scientist Alfred Wegener (1880–1930) proposed the theory of continental drift. Wegener noticed that the shapes of some of the continents looked as if they fit together, like a jigsaw puzzle, suggesting the possibility that they had broken off a single continent and drifted away. Geological changes of this magnitude, if they occurred, could potentially shift the ocean floor. But most scientists at the time could not imagine how such massive structures could move, so Wegener's ideas were not widely accepted.

Another clue came in the 1920s, when the German vessel *Meteor* made one of the first systematic echo soundings. During an expedition from 1925 to 1927, *Meteor* and its crew sailed back and forth across the southern portion of the Atlantic Ocean more than a dozen times, taking some 67,000 soundings and collecting information on salinity and temperature. They identified a long ridge running along the middle of the Atlantic Ocean, which later became known as the Mid-Atlantic Ridge.

But the Mid-Atlantic Ridge was not the end of the story. Danish and British expeditions observed a similar ridge in the Indian Ocean, and during World War II (1939–45), submarine and antisubmarine technologies fueled a more detailed mapping of the ocean floor. Ridges are prominent features throughout the oceans, and in the 1950s, Bruce Heezen (1924–77) at Columbia University in New York collected the soundings data and realized that the ridges formed part of a worldwide system that meanders around the floor of oceans (which, although individually named, are connected). Heezen and cartographer Marie Tharp (1920–2006) produced global maps of what became known as the mid-ocean ridge, although the ridge is centrally located only in part of the Atlantic Ocean. Part (a) of the figure shows a map of the mid-ocean ridge system. Different sections of the ridge are interconnected, although they have separate names, such as the Mid-Atlantic Ridge in the Atlantic Ocean, and the East Pacific Rise in the Pacific Ocean.

The mid-ocean ridge is the longest structure on Earth, extending 45,000 miles (72,000 km) along the ocean basin. Its width varies, averaging about 620 miles (1,000 km), as does its height, which on average rises 1.5 miles (2.4 km) above the surrounding seabed. The mid-ocean ridge takes up more than 20 percent of Earth's surface! Part (b) of the figure illustrates a typical cross-section of the mid-ocean ridge. In the central region of the ridge is a valley, known as a *rift valley*. The valley is usually about 15–30 miles (24–48 km) wide, with a depth of up to two miles (3.2 km). Prominent fissures run perpendicular to the valley, illustrated in part (a) of the figure by cross-hatching.

One of the most interesting features of the mid-ocean ridge is that it is not always found in the ocean! The ridge is sometimes high enough to rise above sea level, and part of its winding course takes it across Iceland, an island in the Atlantic Ocean that straddles the Mid-Atlantic

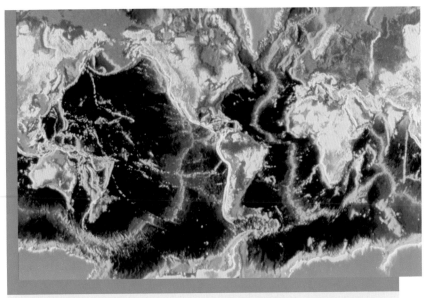

Terrain chart shows the deep ocean basins in black or dark blue and the mid-ocean ridges in light blue. *[OAR/National Undersea Research Program [NURP]]*

Ridge. Iceland is a geologically active island with many volcanoes, and part of the reason is the volcanically active ridge that runs through it. The volcanism of the mid-ocean ridge is a phenomenon that scientists began to understand when they discovered the existence of Earth's tectonic plates.

SEAFLOOR SPREADING

In the 1960s, Canadian researcher J. Tuzo Wilson (1908–93), American scientists Robert S. Dietz (1914–95) and Harry Hess (1906–69), British scientists Drummond Matthews (1931–97) and Frederick Vine (1939–   ), and their colleagues expanded Wegener's earlier ideas. But instead of imagining continents that drift, these researchers realized that Earth's surface is composed of large, rigid plates called tectonic plates. A plate may hold a continent and continental shelf, an ocean basin, islands, or some combination thereof. There are about a dozen major plates, and a few times that number of smaller plates. The plates extend to a depth of about 62 miles (100 km). A map of the main plates is shown in the figure. By comparing the plate boundaries of this map with the location of the mid-ocean ridge, illustrated in the figure on page 36, the ridge corresponds with certain plate boundaries.

Hess was the first researcher to understand what was happening at the plate boundaries that form the mid-ocean ridge. Plates move at various rates, usually in the range of 1–6 inches (2.5–15 cm) per year. Although this may not seem very fast, the movement is enough to make a big difference over time—a plate can travel nearly the length of two football fields in a millennium (1,000 years). And since the plates do not all move in the same direction and at the same speed, they collide, grind against one another, or spread apart. When plates collide, one plate dives beneath another, and this plate usually consists of ocean floor, since oceanic crust is denser than continental crust (which, being lighter, "floats" above sea level). Hess correctly proposed that the mid-ocean ridge is the site of diverging plates—plates that are moving apart—leaving a gap, or rift valley, between them.

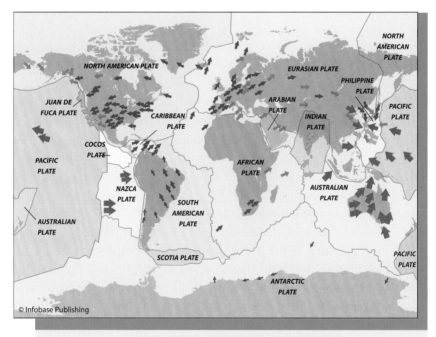

Arrows indicate the direction of movement for the major plates—the size and color of the arrow correspond to the plate's speed, with arrows signifying a higher speed.

Underwater volcanic activity off the coast of Hawaii *(OAR/National Under-sea Research Program [NURP]; University of Hawaii)*

The gap reduces pressure on the rocks below, since it removes the weight of the overlying crust. This reduction in pressure tends to cause rocks to melt; low pressure decreases a material's melting point because the atoms and molecules of the material can separate more easily, and require less heat energy (and thus a lower temperature) to begin to flow. (High pressure raises a material's melting point for the opposite reason.) As plates separate, magma rises to fill the gap. Some of this magma "freezes" or solidifies as it comes into contact with the cool rocks above, and forms new oceanic crust without erupting, at least in the sense of the violent emissions of many volcanoes on land.

Even when magma erupts at the mid-ocean ridge, the cold water and high pressure of the ocean depths preclude the sort of dramatic flows often seen on land. Magma flows in tubes below the seafloor, spilling out of openings into the ocean and chilling quickly, often into rounded, smooth shapes that resemble pillows. Once magma erupts, it is known as lava. The round lava from undersea volcanoes is often referred to as pillow lava.

Although the lava flows of these undersea volcanic events are clearly visible to submersible passengers or video captured from remotely operated vehicles, scientists as yet know few details of the process. Accessible land seismometers are too far away to detect the faint vibrations associated with this volcanic activity. Oregon State University researcher William W. Chadwick, Jr., writing in the December 22, 2006, issue of *Science*, noted that, "Before 1990, not a single eruption was documented on the mid-ocean ridge system, even though many probably occur each year."

Sensors immersed in water or attached to the ocean floor have since occasionally measured the small seismic waves created during lava flows at the mid-ocean ridge, but a lucky break occurred in 2006. Maya Tolstoy, a researcher at Columbia University, had been monitoring a submarine volcano located on the East Pacific Rise for a few years when an eruption suddenly occurred. The unfortunate aspect of the eruption is that many of her team's instruments were now encased in lava. But some of the devices survived, giving Tolstoy and her colleagues a golden opportunity to study a mid-ocean ridge eruption in the act.

Seismic waves are vibrations that travel through rocks (and sometimes fluids), created when a disturbance such as an earthquake suddenly sets into motion a mass of rock, which jostles its neighbors, and so on, propagating the disturbance as a traveling wave. Volcanic eruptions generate waves as the magma flows through chambers, often breaking the surrounding rocks. Tolstoy and her colleagues discovered that seismic activity in the region began to increase starting in October 2003, more than two years before the eruption. The event was intense but lasted only about six hours, and occurred on January 22, 2006. Afterward, the seismic activity decreased quickly. The researchers hope that further study of the elevated activity preceding this event may lead to predictions for other undersea volcanic activity. Tolstoy and her team published their findings, "A Sea-Floor Spreading Event Captured by Seismometers," in a 2006 issue of *Science*.

Although studies of mid-ocean ridge volcanoes are just beginning, the ongoing action of seafloor spreading is indicated by the age of ocean crust. The age of rocks forming the bed of the ocean can be measured by radioactive atoms as well as their relation to certain events such as alterations in Earth's magnetic field (as described in the section "Oceanic Crust and Earth's Magnetic Field," later in this chapter). Crust near the rift is young, and the newly formed rocks become attached to one of the diverging plates. As the plates slowly move, newer crust forms around the edges, and so the age of the rocks gets progressively older with distance from the rift—the rocks located some distance from the rift were formed much longer ago than the new rocks at the rift itself. Most of the ocean crust is less than a few hundred million years old, far less than the age of the Earth. New ocean floor is constantly created at the mid-ocean ridge, and old portions of the floor melt in zones where the ocean floor of a colliding plate dives beneath the other. The oceanic crust is constantly recycled in this process.

FORMATION OF RIDGES

Ridges form where plates diverge and magma wells up through the seam, solidifying into new rocks that rise above the floor, and also spread out as the plates continue to separate. Tectonic plate movement is not well understood as yet, although geologists and marine scientists believe that the plates slide on top of a partially molten layer known as the asthenosphere, which contains flows and currents generated by temperature differences.

Although the ridge has the same basic topography throughout, as shown in the figure on page 36, there is considerable variation along parts of its length. For example, the East Pacific Rise has a gentle slope, and a central valley that is scarcely noticeable in some regions. The Mid-Atlantic Ridge, however, is steep and craggy. Differences in topography are reflected in the names of these two prominent features—a "rise" generally refers to a much more gentle slope than a rugged "ridge." (Even so, the East Pacific Rise is part of the mid-ocean ridge.)

The shape of the mid-ocean ridge depends on the nature of the volcanic activity, but also to a certain extent on the speed of separation of the plates. At the East Pacific Rise, created by the divergence of the Pacific Plate on the western side of the rise and the Nazca Plate on the eastern side, the plates move apart at a rate as high as 6 inches (15 cm) a year. (The Nazca Plate gets its name from a region in Peru.) This fast speed spreads out the new crust, resulting in shallow slopes. At the southern portion of the Mid-Atlantic Ridge, however, the South American and African Plates generally move apart much more slowly, at a rate of about 1 inch (2.5 cm) per year. Rocks pile up as a result, making the Mid-Atlantic Ridge in this region steep and rugged.

Scientists have recently studied even slower separations called ultra-slow spreading. In 2003, Henry J. B. Dick, Jian Lin, and Hans Schouten, all from the Woods Hole Oceanographic Institution (WHOI), published a paper in *Nature* arguing that ultra-slow-spreading portions of the ridge belong to a different category than fast or slow spreaders. In the paper, "An Ultraslow-Spreading Class of Ocean Ridge," the researchers noted that the thin crust at these sections of the ridge result in topographies and heat flows that strongly affect the crust production, "creating a class of ridges that is distinct from previous categories."

The primary example of an ultra-slow spreader is the Gakkel Ridge, a 1,100-mile (1,800-km) section of the mid-ocean ridge located in the Arctic Ocean. (The Gakkel Ridge is named in honor of Soviet Union geographer Yakov Yakovlevich Gakkel.) This ridge spreads by only 0.3–0.5 inches (0.8–1.3 cm) per year, less than half the rate of the "slow" Mid-Atlantic Ridge. Ultra-slow spreading ridges have deep rift valleys but only limited and sporadic volcanic activity; the deep fissures are often filled with cold rock, rather than magma, edging up from below.

Much of the data that Dick and his colleagues used to determine the properties of ultra-slow-spreading ridges come from expeditions sponsored by WHOI and similar organizations. WHOI owns and operates *Alvin*—built in 1964 with funds from the Office of Naval Research—the deep-sea submersible involved in the initial exploration of hydrothermal vents as well as many other projects. The following sidebar provides additional information about WHOI and their important research activities.

Another difference between ultra-slow spreaders and their faster counterparts involves faults. A fault is a crack or fracture in Earth's surface, such as the fractures that occur along plate boundaries. Along most mid-ocean ridges, the rift valley does not line up but is instead divided into segments that are slightly offset. Faults or fracture zones create these offsets; the faults are probably due to weakness in the rocks and the varying rates of spreading. Offsets occur in the Mid-Atlantic Ridge, for example, about every 35 miles (56 km). But the Gakkel Ridge lacks these faults.

The structure and geology of the mid-ocean ridge and its various sections are complicated. And discoveries are just beginning—a great deal of ocean terrain has yet to be mapped in detail. In 2008, for instance, a team of Australian and American scientists aboard the RV *Southern Surveyor,* led by Australian National University researcher Richard Arculus, found a new area of undersea volcanoes and spreading ridges in the Pacific Ocean, northeast of Fiji. Future expeditions will continue to fill in the map.

Woods Hole Oceanographic Institution

Woods Hole is a small village in Massachusetts, located in the southwestern portion of Cape Cod. A salty seaport, Woods Hole has become strongly associated with marine science due to the presence of several important institutions. In 1888, the Marine Biological Laboratory, the oldest marine laboratory in the country, was established at Woods Hole to conduct research and train young scientists. Later, in 1927, a National Academy of Sciences committee formed to determine the best ways to advance the study of oceanography in the United States. The committee, chaired by Frank R. Lillie (1870–1947), then-director of the Marine Biological Laboratory, recommended the establishment of an institution on the east coast to balance the oceanographic research of west coast institutions such as the Scripps Institution of Oceanography in California. The site of this new institution was Woods Hole, and the Woods Hole Oceanographic Institution was founded in 1930.

Woods Hole proved to be an excellent choice. In addition to the proximity of an established marine laboratory, the vil-

EFFECTS OF RIDGES

Ridges are extremely important to study since they are the sites of sea-floor creation and spreading, which is part of the recycling process of oceanic crust. But such a high, long geological feature as the mid-ocean ridge might be expected to have considerable effects on other properties of the ocean, such as the circulation of water. Marine scientists need to understand these effects in order to gain a better understanding of how the ocean works.

Of particular importance is the role of the ocean in weather and climate. The massive body of water comprising the world's oceans exerts a powerful influence due to its many interactions with the atmosphere.

lage has a deep harbor that can accommodate oceanographic research vessels and is close to universities such as Massachusetts Institute of Technology and Harvard University. Today WHOI employs about 1,000 people, including marine scientists, technicians, and vessel crews, and trains a small number of students in a joint degree program with Massachusetts Institute of Technology. The majority of the institution's research budget comes from grants from the National Science Foundation, a United States government agency that heavily invests in research and development. Additional funding sources include the United States Navy, the National Oceanic and Atmospheric Administration, and private donations.

Besides the submersible *Alvin,* WHOI operates three large research ships—*Knorr, Atlantis,* and *Oceanus*—and a smaller vessel, *Tioga.* Researchers at WHOI study a variety of issues concerning the ocean, such as marine biology, undersea chemistry, the ocean floor and sediments, ocean currents, and the relationship between ocean and climate. Some of the exciting projects in which WHOI researchers have participated include finding and filming the wreck of RMS *Titanic,* and the initial explorations of mid-ocean ridges and hydrothermal vents.

Water evaporation at the ocean surface is a critical component of the water cycle—water evaporates from the sea, drifts over land as water vapor, condenses and forms clouds in the atmosphere, falls as precipitation, and then returns to the sea again by way of rivers and streams.

Oceans also affect climates because water has such a great capacity to store heat. Sweating in humans, for example, is a cooling mechanism that uses evaporation to extract a large amount of heat from the skin, thereby cooling the body. Another example is the circulation of cool water to transport heat away from hot machines or motors, which is critical in the operation of automobile engines and other heat-generating combustion engines. A similar process of heat transport occurs in the Gulf Stream, a large current in the Atlantic Ocean that flows from the Gulf of Mexico to northern Europe. One of the main effects of the Gulf Stream is the delivery of heat to northern latitudes; the warm water of the Gulf Stream, along with winds and air flows, moderates the climate of the United Kingdom, which consequently does not experience the severe winters of other places at comparable northern latitudes.

The interaction between ocean and climate can be dramatic and stormy as well. Periodic changes in the temperature of water off the west coast of South America have been associated with storms and droughts over widespread areas across the world. These periodic fluctuations, often called El Niño or La Niña depending on the direction of the change, are the subject of chapter 5.

Maintaining stable ocean temperatures relies on flowing currents and the mixing of water. Water is not a good conductor of heat, which means heat does not travel very easily in water; the reason water is an effective heat transporter is that, as a fluid, it can flow, carrying the heat to other areas. One of the main temperature gradients in the ocean is vertical, varying with depth—sunlight warms the water near the surface, but since the upper layers of the ocean absorb the rays, deeper water is dark and cold. In between is a thin layer of water known as the *thermocline* in which the temperature changes rapidly, going from warm on top to cold on the bottom.

But the presence of the tall, broad mid-ocean ridge can affect this process. Louis C. St. Laurent at Florida State University and Andreas M. Thurnherr at Lamont-Doherty Earth Observatory in New York recently studied mixing of thermocline water around the top of the Mid-Atlantic Ridge. "We know that the mixing of warm surface water with very cold deep water is one of several factors that influence the Earth's climate," St. Laurent said in a press release issued by Florida State University on August 10, 2007. Using special sensors lowered underwater that measure the agitation of water, St. Laurent and Thurnherr discovered that the flow of water through channels and gullies in the ridge creates a lot of mixing opportunities. This mixing, the researchers argue, helps the heat to flow deeper into the ocean, resulting in greater balance. This balance, in turn, may exert effects on currents such as the Gulf Stream. Considering the effects of these currents on weather and climate, any changes in ridge topography wrought by earthquakes or other natural processes could be significant in unexpected ways.

By participating in the recycling of ocean floor, volcanic activity, and the flowing and mixing of waters, ridges and ocean floor spreading are involved in profound as well as subtle ways of shaping the ocean and its properties. The new crust associated with ocean floor spreading is also highly interesting to scientists for another reason—it contains a record of Earth's magnetism.

OCEANIC CRUST AND EARTH'S MAGNETIC FIELD

Sailors use compasses for navigation—or at least they did before the advent of satellites and the global positioning system—because Earth's magnetic field aligns the needle of a compass along a north-south direction. The field extends out into space as well, and when instruments measure the magnetic field in the vicinity of Earth at various points, the measurements are similar to those that would be produced by a giant bar magnet located inside the planet, oriented at a small angle to the axis of rotation. One pole, or end, of the magnet is near the North Pole, and the other is near the South Pole. The locations of the magnetic poles are known as the north magnetic pole and the south magnetic pole.

Earth does not contain a giant bar magnet—and the interior of the planet is so hot that any such magnet would melt—so the field derives from another source. Geologists believe that circulating currents in Earth's iron core generate the magnetic field, but regardless of the source, Earth's magnetism behaves in many ways like a bar magnet. Compasses align themselves to its field, just as iron filings align to the invisible "lines of force" of a bar or horseshoe magnet.

The planet's magnetic field also influences the direction and orientation of magnetic particles in rock. In magma, the particles can move around in the molten rock, and a magnetic field, such as Earth's magnetic field, will align these particles like tiny compass needles. *Igneous rocks—* rocks that solidify from magma or lava—almost always contain some quantity of an iron mineral known as magnetite, which is a natural magnet (hence the name). As the molten rocks solidify, the magnetite particles get frozen into their aligned positions. Here they will stay, reflecting the orientation of Earth's magnetic field when the rock solidified.

An important point to remember is that the spreading plates along the mid-ocean ridge produce new crust. Crust at the rift is fresh, but the crust at some distance away is slightly older because it formed a little while ago and has since moved along with the plate to its new location. The new crust is produced mostly during volcanic activity (except in parts of the ultra-slow spreaders), and the rocks generally contain a lot of magnetic particles. As the new rocks form, the orientation of their magnetic particles adopts the orientation of Earth's magnetic field, which is the predom-

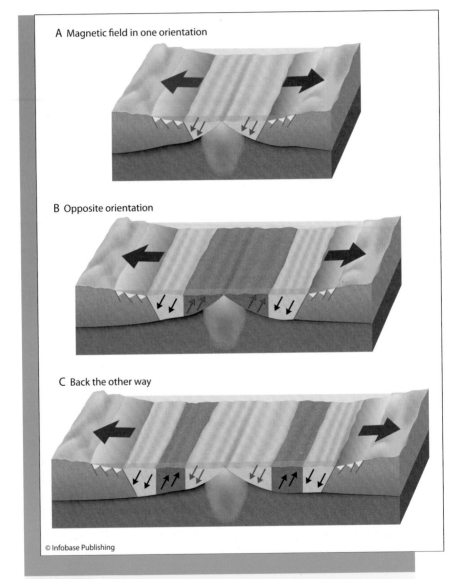

A Magnetic field in one orientation

B Opposite orientation

C Back the other way

© Infobase Publishing

As the seafloor spreads over time, the magnetic field orientation becomes locked in one or the other direction, which is reflected in bands of crust parallel to the mid-ocean ridge.

inant magnetic influence. The ocean floor around the mid-ocean ridge is like a magnetic "tape" recording of the history of the planet's magnetic field, starting at the rift and extending further in time with distance along both sides.

Once aligned, the magnetic particles in these rocks can exert their own magnetic effects. Researchers using sensitive *magnetometers* can determine the orientation of the particles, and marine scientists have sailed vessels towing magnetometers over the mid-ocean ridge. As the magnetometer sweeps perpendicularly over the ridge, it reveals the orientation of the rocks. If Earth's magnetic field has stayed in the same orientation that it has now, the orientation of the magnetic particles in the rock would have stayed the same throughout history. But this is not the case—at various points in time, the poles of Earth's magnetic field have reversed.

Evidence that Earth's magnetic poles had switched also comes from several other sources, including igneous rocks on land. During certain periods, the north and south magnetic poles have been as they are today, and in other periods, the north pole and the south pole of the "bar magnet" changed places, as if someone had flipped over the magnet. In 1963, Frederick Vine and Drummond Matthews realized that the magnetic orientations of strips of oceanic crust around the mid-ocean ridge provide an excellent record of these periods, as indicated in the figure. In the early 1960s, this region was known for its magnetic "anomalies," since no one knew why the magnetism in this area was so variable. The discovery of Vine and Matthews strongly supported the notion of ocean floor spreading, which was a new idea at the time, and helped cement its acceptance.

Reversals occur irregularly, and the time between reversals varies by quite a lot. Some reversals have occurred within 10,000 years, while there have been periods in Earth's history lasting millions of years in which no reversals took place. The last magnetic pole reversal seems to have occurred about 780,000 years ago.

Magnetic pole reversals are important to geologists who are studying the properties and origin of Earth's magnetic field, but these events would also create magnetic disturbances that would probably have some impact on life on the planet, as well as on the electric and magnetic instruments used in modern technology, although no one is sure how significant the impact would be. Researchers have yet to see a pattern or develop a method to predict when the next reversal will happen, assuming another one does (which is likely, given the many reversals in the past).

Careful measurements over the past 150 years indicate that Earth's magnetic field has been losing strength, although the reduction is only a few percent. This decrement may be a prelude to the next reversal, but it can also be a normal fluctuation. Poles of Earth's "magnet" also vary in location—the north and south magnetic poles are not fixed but move around from time to time.

If a reversal is in progress, how long would it take? The magnetic record written in rocks is not always clear, since erosion, plate movements, earthquakes, and other activity jostle the rocks. Reversals may require a few thousand years or so to complete, with the magnetic field losing most of its strength during short phases of the process. But Robert S. Coe of the University of California, Santa Cruz, and Michel Prévot and Pierre Camps of the Université de Montpellier in France have found evidence at a site in Oregon that Earth's magnetic field can change quickly. Using the magnetic orientations of lava flows, the researchers wrote in a report, "New Evidence for Extraordinarily Rapid Change of the Geomagnetic Field during a Reversal," published in a 2002 issue of *Nature,* that their data "suggest the occurrence of brief episodes of astonishingly rapid field change of six degrees per day." This means that the poles would move so fast that in a few weeks the "north magnetic pole" would be located near the equator!

Precise measurements of Earth's magnetic field as revealed in rocks are difficult, and the findings of Coe and his colleagues could contain errors. If further research supports these results, scientists must determine how Earth's magnetic properties can change at such a remarkable rate. Data from mid-ocean ridges will undoubtedly prove valuable in these ongoing investigations.

HYDROTHERMAL VENTS

Magnetic "anomalies" were not the only surprises in store for marine scientists studying the mid-ocean ridge. The mid-ocean ridge hosts its very own unique ecosystem, fueled by hydrothermal vents.

Submersibles have played an important role in undersea research. Initial attempts to explore the mid-ocean ridge came about as part of a project called the French-American Mid-Ocean Undersea Study, often referred to by its acronym, Project FAMOUS. Taking part in this project was *Alvin,* along with French bathyscaphe *Archimède* and submersible *Cyana.* In the early phases of the project, oceanographers assembled ocean floor maps using sounding equipment from the United States Navy and the Scripps Institution of Oceanography, towed by WHOI's ship *Knorr.*

Writing in 1998 in *Oceanus,* a magazine published by WHOI, researcher Ken C. Macdonald recalls "the hushed amazement aboard the research vessel *Knorr* when we first saw high-resolution, deep-tow depth profiles slowly burned into the paper of our malodorous precision depth recorders." Even the bad smell of the equipment could not distract from the fascinating soundings, which showed a deep rift valley with volcanic hills in the Mid-Atlantic Ridge. "These sonar records," continued Macdonald, "were the base maps for the dive expedition, and a team of geologists was assembled to be the first mid-ocean ridge divers using *Archimède* in the summer of 1973."

Dives made during later expeditions explored other parts of the ridge. Then, in 1977, during a dive at the East Pacific Rise, researchers found a hydrothermal vent. The biggest surprise came when so many living organisms were observed inhabiting this otherwise cold, dark place. Since 1977, hydrothermal vents have been the focus of a great deal of study.

Hydrothermal vents have now been discovered in many parts of the mid-ocean ridge. In the early 1980s, divers from the Canadian American Seamount Expedition made the first submersible dives at the Juan de Fuca Ridge, in the Pacific Ocean off the northwest coast of the United States, and found hydrothermal activity and associated life-forms. In 1985, Peter Rona and his colleagues found vents along the Mid-Atlantic Ridge, and vents at several other sites along this ridge have been found since then. A team of Japanese scientists using the *Kaiko,* a remotely operated vehicle, discovered a hydrothermal field in the Indian Ocean in 2000. Hydrothermal vents have even been found in the Arctic Ocean: Researchers from the University of Bergen in Norway and Oregon State University found vents along the Arctic Ridge in 2005. On average, the depth of these hydrothermal vents is about 7,000 feet (2,130 m).

One of the most important characteristics of a hydrothermal vent is the temperature of the water issuing from it. The boiling point of water at sea level is 212°F (100°C), but this temperature depends on pressure. Water molecules that are under little pressure, such as in a pan of water sitting in the thin air on a mountain top, require less energy to break apart,

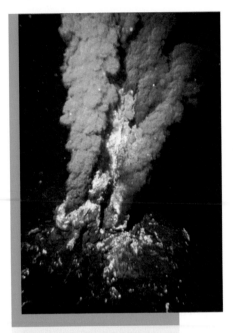

Hydrothermal vent *(OAR/ National Undersea Research Program [NURP]; NOAA)*

so water at low pressure boils at lower temperatures. For example, the boiling point of water at 29,000 feet (8,840 m), which is roughly the height of Mount Everest, is about 158°F (70°C). At high pressure, the opposite occurs, and more energy is needed to wrest a water molecule away from its neighbors. In the depths of the ocean, the boiling point of water is much higher, and liquid water can exist at high temperatures. Water coming from hot water vents can be as hot as 750°F (400°C)—at one atmosphere of pressure, this temperature would be high enough to melt lead.

The temperature of the water strongly affects the appearance of a hydrothermal vent. Vents that issue water having a temperature between 86°F (30°C) and 662°F (350°C) are known as white smokers, since the water has a whitish color. The color comes from compounds of barium, calcium, and other elements that are dissolved in the water. Black smokers emit water that is even hotter, above 662°F (350°C); this hot water contains metal sulfides—compounds of a metal and sulfur, such as iron monosulfide—that are dark.

As the hot water meets the cold water of the ocean depths, the compounds often fall, or precipitate, out of the solution, becoming solid particles. These particles are deposited on the rocks surrounding the vent, gradually building up structures known as chimneys. (Researchers use the term *smokers* to describe hydrothermal vents because the colored water issuing from the chimneys resembles smokestacks of a factory.) Some chimneys can grow 20 feet (6 m) in a year, reaching up to 180 feet (55 m) or higher until they collapse, and start to rebuild.

Where do the dissolved compounds come from? Remember that the mid-ocean ridge is volcanic—magma rises through the fissures of the separating plates. Water from the ocean falls through these cracks, where it encounters molten rock. The water absorbs some heat, raising its temperature, and also picks up some of the minerals in the rocks, which dissolve and flow with the water as it rises to the surface. But the details of this process, and the structure of hydrothermal vents, are not fully known. Researchers are actively exploring this topic, as described later in this section.

Although scientists were not shocked to find hydrothermal vents around the heavily volcanic mid-ocean ridge, the presence of thriving biological communities was unexpected. Nutrients are scarce in the deep sea; sunlight does not penetrate very deeply into the ocean, so the lower layers are cold and dark. On land and in the upper part of the ocean, plants synthesize, or make, carbohydrates by harnessing the energy of sunlight, a process known as photosynthesis. Carbohydrates are rich in energy, and these plants extract this energy to grow and maintain their tissues. Animals do not have the enzymes necessary to perform photosynthesis, but many animals do have enzymes capable of digesting the plants. Photosynthesizing plants form the base of food chains, or webs, in terrestrial as well as upper ocean habitats, by which organisms such as herbivores, which eat plants, and carnivores, which eat other animals, sustain themselves.

Despite the absence of photosynthesizing plants, hydrothermal vent communities thrive in the deep. Some of the most common animals are tube worms. One species, *Riftia pachyptila,* can grow up to 6.5 feet (2 m). The worm, which is red, lives in white tubes made of a substance known as chitin. Under the lights of a submersible, these red worms protrude from their white tubes and, to some researchers, resemble a large tube of lipstick. Other inhabitants include giant clams that grow up to 1 foot (0.3 m) and several varieties of crabs. Several hundred new species have been found at hydrothermal vents.

Forming the base of the food webs at hydrothermal vents are special kinds of microorganisms capable of *chemosynthesis,* extracting energy from chemical compounds in the hot water streaming from the vents. Simple, single-celled organisms known as *archaea* use the high concentrations of hydrogen sulfide (H_2S) in the mineral-rich water as an energy source, rather than sunlight. (These organisms evolved early in Earth's history and are ancient life-forms, accounting for their name, which comes from a Greek term, *archaios,* meaning ancient. Biologists classify them in their own domain, separate from bacteria.) Organisms

Extremophiles—Living in Extreme Conditions

Extremophiles have an affinity to extreme environments. (*Philos* is a Greek term meaning dear or friendly.) Extreme environments that have been found to harbor organisms include highly acidic places such as geysers (and the human stomach), frigid polar regions, and even the interior of certain rocks.

All organisms must have a source of energy in order to maintain life. For archaea and bacteria living around hydrothermal vents, this energy source comes from the chemical bonds in the compounds dissolved in the water. Reactions inside these organisms generate the necessary energy. Hydrogen sulfide, for example, reacts with oxygen to form sulfates and, in the process, releases energy, which the organisms use to drive other reactions to produce sugar (carbohydrates).

A large number of archaea and bacteria species probably exist at hydrothermal vents, as Marine Biological Laboratory researcher Julie A. Huber and her colleagues recently discovered. By analyzing genetic sequences from samples taken from two vents, Huber and her colleagues found thousands of unique sequences, and estimate that as many as 2,700 species of archaea and 37,000 species of bacteria may be present. The researchers published their report, "Microbial Population Structures in the Deep Marine Biosphere," in a 2007 issue of *Science.*

Further study of extremophiles may give astronomers and space explorers a clue about possible life on other planets. Conditions on Mars are cold, dry, and forbidding, but this seemingly inhospitable environment may yet harbor life. Organisms on Earth have found a niche in extreme environments, and simple life-forms may have successfully evolved elsewhere in the solar system.

living in extreme or difficult environments are sometimes called extremophiles, as discussed in the following sidebar.

Some of the larger, more complex organisms of the hydrothermal vent communities feed on these microbes, and so on, up the food chain. But other organisms, such as *Riftia pachyptila,* have an interesting symbiotic relationship with the archaea. A symbiotic relationship benefits both participants, unlike a predatory-prey relationship, which benefits only one. The tube worms have no digestive system, but they allow archaea to live inside their tissues. In return for the home, archaea supply food.

Vents themselves are "energized" by volcanoes, and thrive only as long as the volcanic activity endures. Major geological shifts or disturbances can extinguish a vent. But how do vents form in the first place? Marine scientists used to believe that the presence of magma was sufficient to generate vents, but this may not be the case according to the

Researchers obtain samples from a hydrothermal vent in the Pacific Ocean—layers of bacteria can be seen as a white coating on some of the rocks. *(Pacific Ring of Fire 2004 Expedition. NOAA Office of Ocean Exploration; Dr. Bob Embley, NOAA PMEL, Chief Scientist)*

research of a team of scientists led by University of Oregon geologist Douglas R. Toomey.

Toomey and his colleagues studied the fast-spreading East Pacific Rise with instruments that rested on the bottom and measured seismic waves. These devices produced a detailed image of the activities and geological structure beneath the ocean floor. Sometimes the magma moved laterally, in which case it often cooled and failed to rise to the surface, but sometimes the magma flow was aligned to the fissures between the plates, and the magma pushed upward. The upward rise generated eruptions and enabled the development of hydrothermal vents. In a press release issued by the University of Oregon on March 21, 2007, Toomey said, "What our study shows is that the plumbing system, rather than the amount of available magma, is an important factor that controls eruptions."

But marine scientists still have much to learn about hydrothermal vents. Highlighting this lack of knowledge is "Lost City," a hydrothermal vent field unexpectedly found on December 4, 2000, on a tall undersea mountain in the Atlantic Ocean. This field includes some of the highest chimneys ever found, including one that rises 180 feet (55 m) above the ocean floor. These vents are not located on the Mid-Atlantic Ridge, and their chemistry is different from the vents previously described—they release methane rather than large amounts of hydrogen sulfide and similar compounds—though Lost City also supports a thriving community of organisms. A more thorough understanding of hydrothermal vents and their associated ecosystems must await further exploration.

CONCLUSION

The 45,000-mile (72,000 km) mid-ocean ridge—the largest single volcanic feature on the planet—was not discovered until the 20th century, since most of the ridge is buried beneath a great deal of water. Explorers and marine scientists are just beginning to examine this amazing structure. Researchers have found hydrothermal vents that provide an energy source for unique life-forms and communities, and have used sensitive seismic instruments to study the "plumbing" beneath the ocean floor that sustains magma flows and volcanic activity.

Yet there is much more to learn about mid-ocean ridges and the ecosystems that inhabit them. In some cases, instruments to detect remote events are essential. For example, Toomey and his colleagues positioned instruments on the seafloor to measure seismic waves, taking advantage of this source to provide images of the deep structure under the ridge, as described in their 2007 *Nature* paper. Maya Tolstoy, a researcher whose work on submarine volcanoes is discussed in the section "Seafloor Spreading" earlier in this chapter, has also taken advantage of earthquake waves.

Tolstoy and her colleagues analyzed data from seismic wave detectors that had been placed around a section of the East Pacific Rise in 2003 and 2004. These detectors recorded about 7,000 small earthquakes originating not far beneath the ocean floor. Using sophisticated computer programs, the team of researchers pinpointed the earthquakes, which clustered in a circumscribed region. Tolstoy and her colleagues believe the clustering is because the earthquakes arise from the flow of cold water through the ridge fissures and onto the molten rocks below. The water picks up a lot of heat and seems to move laterally, and then upward, rising through vents located about 1.2 miles (2 km) to the north. As the water absorbs heat from the rocks, the rocks cool slightly, causing them to shrink and crack—which, the researchers believe, is the source of the earthquakes they recorded. The seismic study appears to have outlined the channels under the ridge that feed the hydrothermal vents in the area. This research was reported in a 2008 issue of *Nature,* in a paper titled, "Seismic Identification of Along-Axis Hydrothermal Flow on the East Pacific Rise."

Other research efforts, such as undersea expeditions, gather data directly from the ridge using manned submersibles or remotely operated vehicles. In January 2008, a WHOI expedition sailed in the *Knorr* to the southern Atlantic Ocean, and tested the simultaneous operation of two autonomous underwater vehicles. These robotic vehicles contain their own navigation system and computers that run programs and instructions and are not remotely controlled or attached to the ship. The vehicles explored a little-known area of the Mid-Atlantic Ridge, making sonar maps and sampling chemicals in the water. Researchers on the ship maintained contact with the vehicles using sound waves (radio waves, which are the conventional means of communication, do not travel well through water), and the two vehicles also communicated with one another via a sound-wave link as they explored the ridge. The operation of two vehicles at the same time was successful, and researchers hope the use of multiple vehicles will expand and accelerate undersea research in the future.

As more "eyes" and "ears" explore, record from, and map the mid-ocean ridge, the amount of data will significantly increase. Analyzing huge amounts of data, such as thousands of squiggly lines representing seismic waves or thousands of sequences from the genetic material of hydrothermal vent microorganisms, requires years of work as well as advanced computational and mathematical techniques. But in the end, all the effort is rewarded with a better understanding of one of the last remaining frontiers of Earth—the mid-ocean ridge and the unique inhabitants that call it home.

CHRONOLOGY

1912 c.e.   Alfred Wegener (1880–1930) proposes the theory of continental drift, which, although incorrect in its details, was a prelude to the ideas of plate tectonics and seafloor spreading.

1925–27   Researchers aboard German vessel *Meteor* use sounding equipment to discover the mid-ocean ridge.

1950s   Bruce Heezen (1924–77) and Marie Tharp (1920–2006) produce the first maps of the globe-spanning mid-ocean ridge.

1962   Harry Hess (1906–69) publishes his theory of ocean floor spreading.

1963   Frederick Vine (1939–) and Drummond Matthews (1931–97) discover that the magnetic orientations of strips of oceanic crust around the mid-ocean ridge provide a record of the fluctuations of Earth's magnetic field.

1964   The Office of Naval Research funds the construction of the submersible *Alvin*.

1973   Divers in the French bathyscaphe *Archimède* explore the Mid-Atlantic Ridge.

1977   A team of researchers including Robert Ballard discover the first hydrothermal vent, located on the East Pacific Rise.

1983   Canadian American Seamount Expedition researchers make the first submersible dives at the Juan de Fuca Ridge, and find hydrothermal vents.

1985   Researchers led by Peter Rona discover hydrothermal vents along the Mid-Atlantic Ridge.

2000   A team of Japanese scientists using the *Kaiko,* a remotely operated vehicle, find hydrothermal vents at a portion of the mid-ocean ridge located in the Indian Ocean.

2003   Henry J. B. Dick, Jian Lin, and Hans Schouten argue that ultra-slow spreading portions of the ridge belong to a different category than fast or slow spreaders.

2005   Researchers from the University of Bergen in Norway and Oregon State University discover hydrothermal vents at the Arctic Ridge.

FURTHER RESOURCES
Print and Internet

Chadwick, William W., Jr. "A Submarine Volcano Is Caught in the Act." *Science* 314 (December 22, 2006): 1,887–1,888. The author discusses ocean floor seismometers.

Coe, Robert S., Michel Prévot, and Pierre Camps. "New Evidence for Extraordinarily Rapid Change of the Geomagnetic Field during a Reversal." *Nature* 374 (April 20, 2002): 687–692. Researchers discover evidence at a site in Oregon that Earth's magnetic field can change more rapidly than previously thought.

Dick, Henry J. B., Jian Lin, and Hans Schouten. "An Ultraslow-Spreading Class of Ocean Ridge." *Nature* 426 (November 27, 2003): 405–412. The researchers find another type of ocean ridge.

Fisher, Richard V., Grant Heiken, and Jeffrey B. Hulen. *Volcanoes: Crucibles of Change.* Princeton, N.J.: Princeton University Press, 1997. Written by volcanologists, this book focuses on the geology of volcanoes, eruptions, and emissions, but it includes a brief section on "a little known frontier," undersea volcanic activity.

Florida State University. "FSU Study Finds Deep Ocean Turbulence Has Big Impact on Climate." News release, August 10, 2007. Available online. URL: http://www.fsu.edu/news/2007/08/10/ocean. turbulence/. Accessed June 9, 2009. Describes research indicating that the flow of water through channels and gullies in ocean ridges creates a lot of mixing opportunities.

Huber, Julie A., David B. Mark Welch, Hilary G. Morrison, Susan M. Huse, Phillip R. Neal, David A. Butterfield, et al. "Microbial Population Structures in the Deep Marine Biosphere." *Science* 318 (October 5, 2007): 97–100. By analyzing genetic sequences from samples taken from two hydrothermal vents, the researchers report thousands of unique sequences.

Lahr, John C. "How to Build a Model Illustrating Sea-Floor Spreading and Subduction." Available online. URL: http://pubs.usgs.gov/ of/1999/ofr-99-0132/. Accessed June 9, 2009. This page contains instructions on how to build a model that illustrates the main features of tectonic plate motion and seafloor spreading. Materials needed to construct the model are "a cardboard shoebox, glue, scissors, straight edge, and safety razor blade."

Macdonald, Ken C. "Exploring the Global Mid-Ocean Ridge: A Quarter-Century of Discovery." *Oceanus* 41 (1998). Available online. URL: http://www.whoi.edu/oceanus/viewArticle.do?id=2512&archives=tr ue&sortBy=printed. Accessed June 9, 2009. Macdonald recounts the history of mid-ocean ridge exploration.

Mathez, Edmond A., and James D. Webster. *The Earth Machine: The Science of a Dynamic Planet.* New York: Columbia University Press, 2004. The authors, curators at the American Museum of Natural History, explore the fascinating geological processes that alter and shape the planet. Chapter topics include plate tectonics, lavas, explosive volcanoes, the ocean, and hydrothermal vents.

Scripps Institution of Oceanography and the University of California, San Diego. "Descent to Mid-Atlantic Ridge." Available online. URL: http://earthguide.ucsd.edu/mar/. Accessed June 9, 2009. This Web resource chronicles an exciting expedition to the Mid-Atlantic Ridge made during November 14 to December 14, 2000. Coverage includes an overview of the expedition and a journal describing the highlights.

Tolstoy, M., F. Waldhauser, D. R. Bohnenstiehl, R. T. Weekly, and W.-Y. Kim. "Seismic Identification of Along-Axis Hydrothermal Flow on the East Pacific Rise." *Nature* 451 (January 10, 2008): 181–184. Tolstoy and her colleagues study the cause of undersea tremors around the East Pacific Rise.

Tolstoy, M., J. P. Cowen, E. T. Baker, D. J. Fornari, K. H. Rubin, T. M. Shank, et al. "A Sea-Floor Spreading Event Captured by Seismometers." *Science* 314 (December 22, 2006): 1,920–1,922. The researchers report on the eruption of an undersea volcano.

University of Delaware College of Marine and Earth Studies. "Hydrothermal Vents." Available online. URL: http://www.ocean.udel. edu/extreme2004/geology/hydrothermalvents/. Accessed June 9, 2009. Using an interactive tutorial, a movie, and a recorded interview, this Web resource shows how these "geysers on the seafloor" work.

University of Oregon. "Volcanic Plumbing Dictates Development of Deep-Sea Hydrothermal Vents." News release, March 21, 2007. Available online. URL: http://waddle.uoregon.edu/?id=715. Accessed June 9, 2009. Researchers show that the structure underneath the East Pacific Rise is critical in generating eruptions.

Van Dover, Cindy Lee. *Deep-Ocean Journeys: Discovering New Life at the Bottom of the Sea.* New York: Perseus Publishing, 1995. Van Dover, who has done research at Woods Hole Oceanographic Institution, documents the astonishing discovery of the unusual life and ecosystems near volcanic vents.

Wharton, David A. *Life at the Limits: Organisms in Extreme Environments.* Cambridge: Cambridge University Press, 2002. Life seems to evolve and thrive whenever and wherever it can. This book describes hardy organisms that survive harsh conditions of ice and deserts, as well as at hydrothermal vents deep in the ocean.

Web Sites

Ridge 2000: Venture Deep Ocean. Available online. URL: http://venture deepocean.org/. Accessed June 9, 2009. Supported by funding from the National Science Foundation, Ridge 2000 is a long-term project dedicated to research on mid-ocean ridges. This Web site offers articles about the mid-ocean ridge and its associated hydrothermal vent communities—the feature story in January 2008 was titled "Toasty Tubeworms"—as well as images and research news.

Woods Hole Oceanographic Institution. "Dive and Discover: Hydrothermal Vents." Available online. URL: http://www.divediscover. whoi.edu/vents/. Accessed June 9, 2009. This Web site explores hydrothermal vents, describing the basic geology and the interesting life-forms in its communities. A slideshow of life-forms and vent features, and an animation chronicling the discovery of hydrothermal vents, are included.

———. "Dive and Discover: Mid-Ocean Ridges." Available online. URL: http://www.divediscover.whoi.edu/ridge/. Accessed June 9, 2009. This Web site explains the features of mid-ocean ridges and their geological properties. Included is an interactive illustration showing what a passenger in *Alvin* might see on a typical dive.

3

CREATURES OF THE DEEP SEA

Legends of sea monsters that attack and crush ships have haunted sailors for a long time. Kraken, a legendary sea creature with a bulbous body and writhing tentacles, was believed to lurk in the waters of the northern Atlantic Ocean. Even a formidable battleship such as a man-of-war might have appeared vulnerable to the ferocity of the creature's tentacles. Lookouts in the crow's nest must have kept a wary eye out for these beasts, especially if they had read French author Jules Verne's classic 1870 novel, *Twenty Thousand Leagues Under the Sea,* which portrays a battle between Captain Nemo's crew and a huge squidlike animal.

Fanciful stories of Kraken and other sea monsters began to lose credibility when people starting exploring the ocean depths. Yet the sea is a vast place, and people were not sure what might live beneath the waves. Sometimes animals of monstrous size have washed up on shore. On November 2, 1878, for instance, beach-goers found a dead squid at Thimble Tickle Bay in Newfoundland that reportedly had a tentacle measuring 35 feet (10.7 m) and an eye nearly 20 times as big as the eye of a person! Whalers often observed circular scars around the heads of sperm whales, as if the suction cups of a giant squid's tentacles had gripped the whale during some fierce struggle far below the surface.

With the help of submersibles and remotely operated vehicles, scientists and explorers have extended their range of observation to include the great ocean depths. This frontier of marine science includes a wealth of life, and although no Kraken or similar monsters have been exposed, researchers

Colorful marine life in the Pacific Ocean at a depth of 560 feet (170 m) *(Pacific Ring of Fire 2004 Expedition. NOAA Office of Ocean Exploration; Dr. Bob Embley, NOAA PMEL, Chief Scientist)*

have found plenty of other surprises. Chapter 2 described one such discovery—hydrothermal vent communities—and this chapter delves further into the subject. Researchers are studying how animals survive in the deep sea, how they adapt to harsh conditions, and how this knowledge can be applied to solve some of society's most difficult engineering and medical problems.

INTRODUCTION

Many 19th-century scientists dismissed notions of huge sea creatures, and a few scientists even gave little credence to the belief that any creature could live in the deep. For example, Edward Forbes (1815–54), a respected and talented British scientist, concluded that the deep ocean was an azoic zone, meaning that it has no life. (The term *azoic* comes from a Greek term *zōē*, life, and *a*, without.) Forbes based his belief on dredging operations he conducted in the 1830s and 1840s to bring up samples from the ocean's lower regions. Sunlight fails to penetrate the deepest depths of the ocean, and Forbes noted that no photosynthesizing plants live there. Photosynthesis is a plant's way of making food by using the energy of sunlight, and plants were the foundation of the food chain in the ecosystems with which Forbes was familiar. The argument Forbes made was that without photosynthesis there could be no food chain, and therefore no life.

Hydrothermal vent communities get around the lack of photosynthesizing plants with the help of archaea capable of chemosynthesis—producing nutrients from various chemical compounds. But long before researchers discovered hydrothermal vent communities in 1977, scientists had dredged up strange creatures from the bottom of the ocean. Even as early as 1818, British explorer Sir John Ross (1777–1856) discovered worms in the mud samples he retrieved from as deep as one mile (1.6 km) in Baffin Bay, which lies between Canada and Greenland. An "azoic" deep sea did not seem to hold true everywhere. Researchers aboard the HMS *Challenger* during an oceanographic expedition around the world in 1872–76 retrieved many samples from the ocean floor and discovered a variety of deep-sea organisms.

Thanks to the diligent efforts of the voyagers of the HMS *Challenger* and the many researchers who followed them, marine biologists have discovered about 250,000 species that live in the ocean. Some of these animals float, some of them swim, but 98 percent of these species live on the bottom. Marine scientists refer to the bottom of the ocean as the *benthic* environment (from the Greek word *benthos,* which means depth).

The ocean floor is a varied environment, and includes sandy intertidal regions, mid-ocean ridges, ocean basins, and deep trenches. Most benthic species live on the continental shelf, the part of the ocean bordering the continents. (See the figure on page 6.) These bottom-dwellers enjoy a shallow and hospitable, sunlit environment.

Some of the most common benthic organisms are corals. These small creatures capture food with stinging tentacles and live in colonies, where they construct a structure made of calcium carbonate for protection. Over time, these colonies form coral reefs, which are usually found in shallow, tropical water, although a few types of coral can survive in cold, deep water.

Marine scientists divide the sea into various zones known as pelagic zones (from the Greek term *pelagos,* sea), based on depth. The zones are as follows:

- epipelagic (from the Greek word *epi,* "on"): surface to 660 feet (200 m)
- mesopelagic (from the Greek word *mesos,* "middle"): 660–3,280 feet (200–1,000 m)
- bathypelagic (from the Greek word *bathos,* "deep"): 3,280–13,120 feet (1,000–4,000 m)
- abyssopelagic (from the Greek word *abyssos,* "without bottom"): below 13,120 feet (4000 m)

Benthic environments are also divided into zones, the two deepest being the abyssal zone, which refers to the ocean floor at depths between 13,120 feet (4,000 m) and 19,680 feet (6,000 m), and the hadal zone, which refers to the ocean floor at any greater depth. (The term *hadal* derives from *Hades,* which in Greek mythology is the underground residence of the dead.)

Oceanic environments can also be divided into zones based on the penetration of sunlight. In the euphotic zone, there is enough sunlight to support photosynthesis; this zone begins at the surface and usually does not extend deeper than 330 feet (100 m). (The term *euphotic* comes from the Greek words *eu,* "good," and *photos,* "light.") Only a small amount of light is available in the disphotic zone (from the Latin *dis,* "separate" or "apart"), which is located between 330 feet (100 m) and about 3,300 feet (1,000 m). Deeper regions comprise the aphotic zone, where light fails to penetrate.

The exact depth that sunlight can penetrate depends on the water's *turbidity*—the amount of particles suspended in the water, which tend to reflect sunlight. Another factor is the number of organisms known as *plankton.* Plankton are organisms such as tiny algae that float or drift in the ocean. (The term *plankton* comes from a Greek work, *planktos,* which means "drifting.") Although normally these organisms are quite small, they are so common that the majority of the ocean's biomass—the mass of all organisms—is from plankton. (And since the oceans are so large compared to the land, plankton also make up the majority of Earth's biomass.) Plankton block or absorb sunlight—many engage in photosynthesis—and help keep the deep layers of the sea in the dark.

A quick way to estimate water's transparency is to use a simple device known as a Secchi disk. Named for its inventor, Italian scientist Pietro Angelo Secchi (1818–78), the disk is 8–16 inches (20–40 cm) in diameter, painted white or black and white, and is attached to a rope. A researcher lowers the Secchi disk into the water until it disappears from view. The depth at which it disappears is related to the turbidity of the water, which prevents light from reaching and reflecting off the disk, thereby obscuring it from view.

In addition to the absence of sunlight, the inhospitable environment of ocean's depths includes low temperatures and high pressure. Light and warmth fail to reach far into the ocean, and the weight of the water exerts tremendous pressure on undersea objects. The temperature of water at a depth of about one mile (1.6 km) or lower remains about 37.4°F (3°C) throughout the year at all latitudes. Pressure increases at a rate of one atmosphere for every 33 feet (10 m), so that at a depth of 33 feet (10 m), the pressure is equal to twice what a person experiences at ground level. (At a depth of 33 feet [10 m], a diver feels the pressure of the atmosphere plus the pressure from the water.) At 330 feet (100 m), the pressure is 11 times greater than ground level, and is equal to 162 pounds (73.6 kg) per square inch (6.25 cm²).

Photosynthesizing plants emit oxygen, and their abundance in the euphotic zone results in plentiful oxygen in the upper level of the ocean. Fish breathe by extracting this dissolved oxygen in their gills. The lack of sunlight in deeper layers precludes photosynthesis, and the oxygen level drops precipitously in the disphotic zone. Although parts of the ocean have little oxygen, much of the deeper regions is oxygen-rich, despite the absence of photosynthesis. The oxygen comes from oceanic circulation, especially from cold, well-oxygenated water from the polar regions. This oxygen means that fish and other marine life can live deep in the ocean, even in the deepest hadal zone, provided the animals can handle the cold temperature and extreme pressure—and can find something to eat.

ANIMALS OF THE DEEP

Photosynthesizing plants form the base of the food chain for most ecosystems on land, and the absence of these organisms in the deep parts of the ocean means that the food chain of these communities must be different. One of the main consequences of a lack of plants is that many deep sea creatures have nothing to eat but each other. And they do so, resulting in a "Wild West" sort of habitat in which the law is eat or be eaten. Deep-sea animals tend to look fearsome, with large jaws and long teeth, which suits their behavior.

Some species use speed or brute force to capture their prey, while others resort to trickery. One way to attract a potential meal is to emit light, which stands out in the otherwise total darkness of the deep and may mimic food, luring hungry creatures. *Bioluminescence* is the process by which organisms emit light, and such organisms are called bioluminescent. Light is also useful for attracting mates, which greatly increases reproductive opportunities. Without some way of signaling their presence, animals would have to rely on chance encounters, bumping into a receptive mate by accident—but bumping into objects in the deep is risky, since on most occasions the other object is hungry. Bioluminescence is so useful and common that many deep sea creatures have vision, despite the darkness of their environment. To gather as much light as possible, their eyes tend to be large or especially sensitive. The way in which organisms generate light is discussed in a later section, "Bioluminescence."

Sloane's Viperfish, a deepwater predator *(Dante Fenolio/ Photo Researchers, Inc.)*

Steven H. D. Haddock of the Monterey Bay Aquarium Research Institute (MBARI) in California and his colleagues recently discovered an animal known as a siphonophore that "fishes" with a red bioluminescent lure. Siphonophores live in colonies, and they feed by stinging their prey with tentacles, similar to jellyfish. Most siphonophores are bioluminescent, but Haddock and his colleagues discovered one that uses a bioluminescent appendage in an unusual way. As reported in "Bioluminescent and Red-Fluorescent Lures in a Deep-Sea Siphonophore," published in a 2005 issue of *Science,* the researchers discovered siphonophores that waved and flicked luminous filaments amid the stinging tenta-

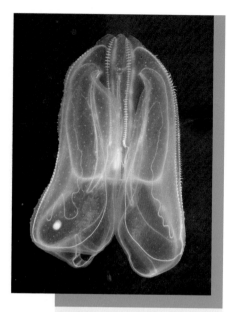

Bioluminescent marine organism called a comb jelly *(Gregory G. Dimijian/ Photo Researchers, Inc.)*

cles. Fluorescence, a process by which objects absorb and reemit light, also contributed to the glow. Haddock and his colleagues believe the lights entice fish, which are the prey for these particular siphonophores. The animals were discovered at depths of 5,250–7,540 feet (1,600–2,300 m), where fish are scarce, so these siphonophores need to use lures to attract food.

To find and study creatures such as these deep sea siphonophores, Haddock and his colleagues logged many hours at the controls of remotely operated vehicles (ROVs). Visiting the deep environment is a must, for as discussed in the next section, deep-sea creatures do not fare well when brought up to the surface. Research institutions such as the Marine Biological Laboratory (MBL), Woods Hole Oceanographic Institution (WHOI), and MBARI employ submersibles and ROVs to study animals living in the ocean depths. WHOI was discussed in a sidebar on page 44, and the following sidebar provides some more information on MBARI.

Monterey Bay Aquarium Research Institute (MBARI)

Located at Moss Landing, California, on the shore of Monterey Bay less than 100 miles (160 km) south of San Francisco, MBARI is a relative newcomer in oceanography research. David Packard (1912–96) established the not-for-profit research institute in 1987, a few years after helping his daughters develop the well-stocked Monterey Bay Aquarium. Packard, an engineer, is also well known for his role in founding the computer company Hewlett-Packard, or HP as the company is often called, along with William Hewlett (1913–2001). With a strong conviction in the power of both science and engineering, Packard wanted to start an oceanographic institute that emphasized the combined efforts of scientists and engineers.

Monterey Bay is an excellent location for oceanographers, partly for the same reason that San Francisco is a good choice for geologists—it is near a geologically active plate boundary, where the Pacific Plate is grinding past the North American Plate. The Bay is home to a diverse selection of marine species, and also has a deep submarine canyon, known as the Monterey Canyon. The canyon begins near Moss Landing, and stretches nearly 100 miles (160 km), with the bottom reaching a depth of up to about 13,120 feet (4,000 m). MBARI oceanographers do not have to sail very far to reach deep water!

From the beginning, MBARI focused on ROVs, advanced instruments, and computer technology in their research objectives. A year after the institute was established, researchers made their first scientific expedition with a research vessel *Point Lobos* (converted from an oil-rig supply boat) and an ROV, the *Ventana*. MBARI has since grown to a staff of about 200 workers who conduct research on the biology and chemistry of benthic environments, as well as many other projects.

The ocean depths are not extensively mapped, and the biological environment of the deep sea is very much a frontier of science as well. New discoveries are common on almost every expedition. In 2002–05, an international team of researchers conducted a series of expeditions in the Weddell Sea, off the coast of Antarctica, and nearby regions, using the German research vessel *Polarstern.* Scientists collected specimens from a depth range of 2,450–20,820 feet (748–6,348 m). The expeditions were part of ANDEEP, Antarctic benthic deep-sea biodiversity project, involving researchers from 17 institutions and organizations. Among the specimens were hundreds of new species; for example, out of 674 species of isopod (an order of crustaceans), 585 had not been seen before. The team of researchers published their report, "First Insights into the Biodiversity and Biogeography of the Southern Ocean Deep Sea," in a 2007 issue of *Nature.*

Discoveries will undoubtedly continue as researchers explore the deep sea. According to the MBARI Web site in June 2009, "Almost every time MBARI midwater researchers make ROV dives more than a mile below the surface, they see animals that have yet to be given scientific names." Learning about these new species is imperative in order to understand the deep-sea ecosystems.

ADAPTING TO THE DEPTHS

One of the most important attributes of deep-sea animals is their ability to withstand the tremendous pressure of the ocean depths. Since pressure increases by the equivalent of one atmosphere for every 33 feet (10 m), the pressure is 1,485 pounds (675 kg) per square inch (6.25 cm^2) at 3,300 feet (1,000 m) below the surface. This is enough force to crush the hulls of submarines.

The problem for organisms or objects that venture into deep water is unbalanced pressure. In a submarine, for instance, the pressure inside the vessel is kept at about one atmosphere so that its sailors can breathe normally. On the other side of the hull is the crushing force of the water. This means that a great force is squeezing the submarine's hull from the outside, and there is little force from the inside to counterbalance it. Even steel cannot withstand too much squeezing, so submarines do not dive past certain depths.

Humans and other organisms that usually live on or around Earth's surface experience a similar situation if exposed to tremendous

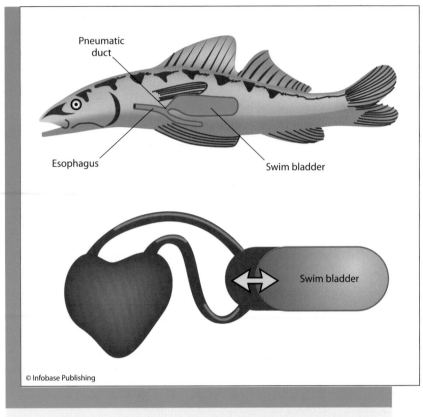

Pneumatic
duct

Esophagus

Swim bladder

Swim bladder

© Infobase Publishing

Swim bladders are sometimes connected to the esophagus, as in the fish shown at the top of the figure, facilitating the flow of air into and out of the bladder. In other cases, the swim bladder exchanges air with the blood, illustrated in the bottom of the figure.

pressure. Although humans are not made of steel, the body is surprisingly durable; about 65 percent of the human body's weight is water, which can withstand such forces (as it does in the ocean), and the bones are also strong. The problem with deep-sea diving is the unbalanced pressure on the vital air pockets of the human body, especially the lungs. Air inside the lungs gets squeezed as a human diver descends, and the lungs would eventually collapse.

Fish are also affected. Objects that are less dense than water tend to float; for instance, submarines dive by flooding tanks with water, which increases their density, and ascend by pumping the water out. Some species of fish maintain air sacs known as swim bladders to make it easier for them to keep from sinking. The figure illustrates an example of a swim bladder and its location in the fish's body. Swim bladders add or remove air from the chamber either by a connection with the esophagus (the tube connecting the mouth and stomach) or by gaseous exchanges with the blood.

As a fish's depth changes, the pressure on any air inside the animal also changes. A swim bladder of a fish deep in the ocean is under a great deal of pressure, and its volume is tiny. If such a fish is brought up to the surface, the tremendous pressure of the air inside the swim bladder would no longer be balanced by the pressure of the water in the depths of the ocean. This imbalance would result in a situation similar to the crushing of a human diver's lungs, but in the opposite direction—the swim bladder swells because its internal pressure is much greater than its surroundings. As a result, the fish dies.

Some deep-sea fish have adapted to their environments by using a substance other than air in their swim bladders. Fat is a common alternative. Organisms store fats, which are energy rich compounds, in certain cells, and fats are part of the diet. At the crushing pressures of the deep ocean, air is so squeezed that its density is close to that of fatty tissue anyway, so fat will work just as well as air. Other deep-sea creatures manage quite well without a swim bladder at all. Bottom-dwellers, for example, have little need for buoyancy. These organisms are not afflicted with bursting swim bladders when researchers diving in submersibles or operating ROVs capture and bring specimens to the surface.

Despite the swim bladder adaptations, all deep-sea animals, including those without swim bladders, usually die when brought to the surface. The death of animals with air-filled swim bladders is easy to explain, but the deaths of other animals are not. This is an active area of marine research.

One explanation involves cell membranes. A cell, the basic unit of life in all organisms, contains nutrients and structures surrounded by a membrane composed mostly of lipids, which are fatty substances. The membrane maintains a controlled environment inside the cell and regulates substances entering and exiting the cell. Cells in multicellular organisms work together to form tissues and organs, but must maintain their membrane integrity in order to function. In deep-sea environments, the intense pressure on a cell's membrane might squeeze the membrane so much as to disrupt the access channels by which molecules move in and out, if the animals did not make some sort of adaptation.

No one is certain what sort of adaptation deep-sea organisms have made in their cellular membranes, but in 1985, Edward F. DeLong and A. Aristides Yayanos of Scripps Institution of Oceanography reported that a certain deep-sea bacterium elevated the amount of unsaturated fatty acids when exposed to greater pressure. Fatty acids consist of chains of bonded carbon atoms with hydrogen atoms attached at various points. They are important components of membrane lipids. The carbon-carbon bonds are single covalent bonds in saturated fatty acids, but unsaturated fatty acids contain some double bonds, meaning that two pairs of electrons are shared. These double bonds cause kinks in the chain, leading to lower melting points and a generally squishier structure. Vegetable oils, for instance, have much unsaturated fat, while solid fats tend to be saturated. By using more unsaturated fatty acids when the pressure increases, bacteria may be preventing their membranes from becoming too rigid. If this is true in general, then a sudden release of this pressure, as when the organism is brought from the deep sea to the surface, would result in cellular membranes that are too fluid and may become "leaky," killing the organism. DeLong and Aristides reported their findings, "Adaptation of the Membrane Lipids of a Deep-Sea Bacterium to Changes in Hydrostatic Pressure," in a 1985 issue of *Science*.

Other researchers are interested in proteins. Proteins, like fats, are an essential component of the diet, but digested proteins are broken down into their component amino acids and later reassembled to make the body's own special proteins. The assembly is based on information contained in genes, which code for the sequence of amino acids to make any given protein the body needs. When the amino acids are assembled in the correct sequence, the protein automatically adopts a three-dimensional shape that is critical for it to carry out its function. Enzymes, for example, are proteins that speed up chemical reactions, and an enzyme usually has a special shape in order for it to bind certain molecules and bring them together, facilitating the reaction.

How do proteins in deep-sea organisms maintain their shape under extreme pressure? Researchers do not know the answer, but Amanda A. Brindley of the University of Kent in the United Kingdom and her colleagues have recently compared an enzyme called lactate dehydrogenase from a deep-sea fish, *Corphaenoides armatus,* with a similar enzyme in another fish, *Gadus morhua,* which lives in shallow water. The researchers sequenced the amino acids of the two enzymes and found only 21 differences, yet the enzyme of the deep-sea fish displayed *hyperbaric* stability—it tended to maintain its shape and function despite high pressure. By using techniques of molecular biology to rearrange the amino acid sequence, the researchers discovered that the most important adaptations of the hyperbaric protein came from one end of the protein. The researchers published their report, "Enzyme Sequence and Its Relationship to Hyperbaric Stability of Artificial and Natural Fish Lactate Dehydrogenases," in a 2008 issue of *PLoS One.*

Further investigations of hyperbaric proteins and how they have adapted to high pressure will increase scientific knowledge of these critical molecules, which is an extremely important field of research in biology and related disciplines. As Brindley and her colleagues write in their report, "An understanding of these adaptations will provide important insights into the structure/function relationships of proteins, particularly in relation to protein stability and enzyme catalysis under the extremes of environmental stress." These insights can have dramatic consequences, as discussed later in the chapter, in the section titled, "Applications of the Biology of Extreme Environments."

Another fascinating issue concerns animals that do not live in the deep sea but occasionally venture into great depths. These animals, such as sperm whales, are remarkable in their ability to adjust temporarily to a hyperbaric environment, yet resurface a short time later. Sperm whales will be discussed in a later section titled, "Whales and Giant Squid."

DEEP-SEA ECOSYSTEMS AND FOOD CHAINS

In addition to the physiological adaptations of deep-sea organisms to their extreme environmental conditions, a critical aspect of life in the depths is the food chain. Some benthic ecosystems have evolved to take advantage of special resources. Hydrothermal vents harbor communities in which tiny organisms such as archaea produce food by chemosynthesis, using minerals and compounds spewing out of hot water vents on the bottom of the ocean.

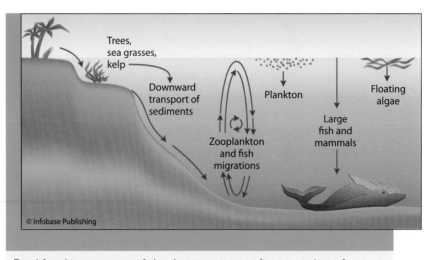

Food for the creatures of the deep sea comes from a variety of sources.

But some deep-sea animals live at a great distance from any hydrothermal vent. What do these animals find to eat? Many of them, as mentioned above, are carnivores. But some form of food production or food input is generally necessary for a stable ecosystem; human civilization, for example, would not survive if there were no farmers, even though most people these days have other jobs. For deep-sea creatures, the food input is either carried by circulating currents or drifts down from above, from the euphotic zone.

At the bottom of the ocean is a sediment layer consisting of mud and biological materials. Detritus—the remains of dead organisms—and waste products fall and settle on the ocean floor, as depicted in the figure. Most of the food production of the euphotic zone is consumed at or near the surface, but "table scraps" are available for scavengers along the bottom of the ocean. It is, one would suspect, a tough way to make a living, but hardy organisms exist in almost every conceivable ecological niche on the planet.

One particularly rich source of food comes from the death of a large animal such as a whale, which then sinks to the bottom. Blue whales, the largest species, can weigh up to 170 tons (155,000 kg). Although most whales are significantly smaller, whale carcasses contribute a large amount of sustenance for deep-sea communities.

How long do whale carcasses last on the ocean floor? Some marine researchers have discovered whale carcasses on the ocean floor and then repeatedly visited them to chart the progress of decay. Other researchers have conducted experiments by sinking newly discovered carcasses at a specific place, and monitoring the results. University of Hawaii researcher Craig R. Smith and Woods Hole Oceanographic Institution researcher Amy R. Baco reviewed the results of these observations in a paper, "Ecology of Whale Falls at the Deep-Sea Floor," published in the 2003 *Oceanography and Marine Biology: An Annual Review.*

According to Smith and Baco, benthic organisms consume whale carcasses in at least three stages. First is "a mobile-scavenger stage lasting months to years, during which aggregations of sleeper sharks, hagfish, rattails and invertebrate scavengers remove whale soft tissue at high rates," as much as 88–132 pounds (40–60 kg) per day. For example, Smith and Baco note that more than 90 percent of the flesh of a 5.5-ton (5,000-kg) whale on the ocean floor near San Diego was gone in four months. Next there is "an enrichment opportunist stage (duration of four months to years) during which organically enriched sediments and exposed bones

Humpback whales *(Dr. Louis M. Herman/NOAA)*

are colonised by dense assemblages" of various organisms. Finally, the third stage is "a sulphophilic ('or sulphur-loving') stage lasting for decades, during which a large, species-rich, trophically complex assemblage lives on the skeleton as it emits sulphide from anaerobic breakdown of bone lipids." (Anaerobic breakdown means that oxygen is not involved.) This stage includes organisms similar to the chemosynthesizers found at hydrothermal vents.

The finding of similar organisms on whale carcasses and at hydrothermal vents has fueled speculation that whale carcasses are important in the evolution of vent organisms. Perhaps the migration of certain creatures from vent to vent depends on the availability of temporary "way stations" composed of whale falls. To investigate the relatedness of organisms found at whale carcasses with those at hydrothermal vents, University of Maine researcher Daniel L. Distel joined Smith, Baco, and other colleagues in analyzing *deoxyribonucleic acid* (DNA) samples.

Using DNA to study the evolution of life and the relationship of diverse species has become a common procedure. DNA is the hereditary material that passes along information essential in the development and basic functions of every organism. Sequences of DNA form genes that code for the organism's specific set of proteins, and each species has its own unique set that has evolved over time. But two closely related species that only recently evolved from a common ancestor share many genes and DNA sequences; for instance, the DNA of humans and chimpanzees differ by only a few percent. By sequencing and analyzing DNA, researchers acquire an important clue on the evolutionary relatedness of different species.

Distel, Smith, Baco, and their colleagues analyzed the DNA of a type of mussel commonly found in whale falls and in sunken wood, such as from old vessels. (A mussel is a clamlike mollusk.) DNA sequences show that these mussels are closely related to a family of mussels found only at hydrothermal vents and other specialized habitats. The researchers believe these organisms belong to the same subfamily of mussels, which evolved and adapted to their peculiar environment. But vents are sparse on the ocean floor, so how do these organisms colonize new territories? Distel, Smith, Baco, and their colleagues propose that these organisms spread across the ocean floor from vent to vent by taking small "steps," which consist of whale falls and sunken wood. The researchers published their report, "Do Mussels Take Wooden Steps to Deep-Sea Vents?" in a 2000 issue of *Nature*.

The unique environments of the deep sea create a lot of interesting properties shared by many of its inhabitants. But one of the most important properties also exists in certain common insects known as fireflies.

BIOLUMINESCENCE

Light is a form of energy, as is heat and electric current. According to the laws of physics, energy cannot be created or destroyed, but one form of energy can be converted into another. In a type of lightbulb known as an incandescent bulb, as Thomas Edison (1847–1931) and other inventors of the 19th century discovered, an electric current passing through a thin filament raises the filament's temperature and causes it to emit light. Electrical energy is converted into heat, and a hot body emits light, a process called incandescence. Unfortunately for consumers who must pay electric utility bills, much more of the energy is turned into heat than light in incandescent lightbulbs, so a lot of the electrical energy is wasted. Fluorescent lamps, which transform ultraviolet light into visible light, do not use incandescence, and they produce much less heat and are more efficient (though more expensive) than incandescent bulbs.

Bioluminescence is an entirely different process. Certain organisms are able to convert chemical energy into light, or, in other words, generate light with special chemical reactions. This light is often called a "cold light," because unlike incandescence, high temperature is not an essential feature—bioluminescence is quite efficient, with only a small amount of the energy in the process ending up as heat. The following sidebar discusses the chemical reaction involved in bioluminescence.

Few land animals employ bioluminescence, although fireflies are well known to do so. Many firefly species emit flashes of light at various rates and durations in order to attract mates. The light usually comes from the insects' abdomen, the tail-end segment. This "cold light" is not harmful because it generates little heat.

Bioluminescence is not the same as phosphorescence. Phosphorescent materials absorb light or other forms of energy, and then later emit this energy as light. Glow-in-the-dark toys or stickers, for example, are usually phosphorescent. Bioluminescence requires chemical reactions involving luciferin, catalyzed by an enzyme, luciferase, as discussed in the sidebar.

The Chemical Reaction of "Cold Light"

Some chemical reactions, known as exothermic reactions, emit heat. For example, combustion—such as a burning match—gives off heat. In combustion, a substance combines with oxygen chemically to form products such as carbon dioxide and water, and some of the potential energy stored in the substance's chemical bonds gets transformed into heat. Some of this potential energy often gets transformed into light as well—the flame of a burning match, for example. In bioluminescence, the chemical energy is mostly converted into light.

Bioluminescence reactions generally occur inside the bodies of organisms and involve several molecules. One molecule, called luciferin, emits light when it reacts with oxygen (O_2). An enzyme, known as luciferase, catalyzes or speeds up this reaction. (Luciferin and luciferase get their name from *lucifer*, a Latin term meaning light-bearer.) There are many different kinds of luciferin and luciferase. Often a molecule of adenosine triphosphate (ATP) is also required. Cells use the chemical energy of ATP to drive a lot of chemical reactions that would not otherwise take place, and this energy is necessary for certain kinds of luciferin and luciferase.

In general, the reaction can be symbolized as follows:

luciferin + O_2 (catalyzed by luciferase) → oxyluciferin + light

Oxyluciferin is a compound of luciferin and oxygen, and will no longer react with oxygen to produce light. In most cases, the luciferin must be replenished. Some animals accomplish this with dietary intake, while other animals engage in chemical reactions that synthesize fresh luciferin from other molecules.

The absence of light in aphotic zones of the deep sea means that animals that live there either generate their own light or do without. Uses of bioluminescence include luring prey, as in the siphonophore that Haddock and his colleagues discovered, as well as attracting mates. Bioluminescent marine organisms also use light to warn off competitors in their territory and confuse an attacker by suddenly flashing lights. Emitting light is so important in the dark environments of the ocean depths that the majority of deep-sea organisms are bioluminescent.

On occasion, bioluminescence is a nuisance. Some deep-sea creatures are transparent and like to remain hidden from predators, but if a transparent animal consumes a bioluminescent meal, it becomes bioluminescent itself, at least for a little while! The unwanted bioluminescence would make the animal an easy prey for larger predators. Many of these transparent animals have opaque (nontransparent) digestive tracts, possibly to avoid this potentially serious problem.

Some bioluminescent organisms create a steady light while others emit flashes. In either case, these lights punctuate an otherwise dark and mysterious environment, signaling fights for mates or food. But other interactions in the ocean depths involve creatures that are only occasional visitors. One of these, the sperm whale, sometimes hunts gigantic prey—the giant squid—which is also one of its most powerful enemies.

WHALES AND GIANT SQUID

Sperm whales are the largest toothed whales (the jaws of other whales have baleen, a bony substance that filters out food such as plankton, rather than teeth). Adult male sperm whales can weigh 50 tons (45,450 kg) and have enormous heads, which contain the largest brain of any animal on Earth. The head also contains a large quantity of a fluid known as spermaceti, which sailors once mistook for sperm, hence the name of the whale. Sperm whales live in all regions of the oceans, typically around deep water. They eat squid, sharks, and fish.

Whales are mammals—they breathe air and must surface often. When a whale dives, it must hold its breath until it can resurface. Sperm whales hunt for food on their dives, and on average spend about half an hour underwater at a time. But deep, prolonged dives may exceed an hour.

How deep can a sperm whale go? No one knows, although sonar operators have detected signs of whales at depths greater than 6,560 feet (2,000 m). Scientists can study the underwater movements of whales by attaching a sensing device coupled with a transmitter to the whale's back. In the wild, this is not easy to do—researchers must get close to the whale and attach an instrument that does not harm the animal or interfere with its behavior. Most researchers use devices having suction cups that stick to the whale's skin; the device is usually designed to pop off after a short period of time, or will come off naturally. The package includes such instruments as a depth gauge and a radio transmitter. Although radio waves do not tend to travel far in water, whales must resurface periodically to breath, during which times the transmitter can readily send the instrument's data to receivers on board vessels or through satellite networks.

In 2002, William A. Watkins of WHOI and his colleagues found some interesting results from tracking sperm whales. Watkins and his associates had earlier attached a radio tag to a 39-foot (12-m) sperm whale in the southeast Caribbean Sea, and monitored the transmissions over a period of four and a half days. The researchers regarded any submergence of the whale that lasted longer than three minutes as a dive, and recorded 158 dives over this period. Of these 158 dives, 65 were shallow—less than 650 feet (200 m)—and 93 were deep, averaging 3,250 feet (990 m) and ranging from 1,380 feet (420 m) to 4,360 feet (1,330 m).

The pressure at 4,360 feet (1,330 m) exceeds 130 times the pressure at sea level. Sperm whales have lungs, so they should be subject to the same difficulties that afflict humans who dive too deeply, yet they manage to survive. Their physiology must be adapted to withstand these great pressures for short periods of time.

One of these adaptations involves allowing the lungs to collapse without injury. If a human lung is subjected to a great deal of pressure and becomes squeezed too tightly, blood vessels in the lung will break and start to bleed, filling the cavity with a suffocating amount of blood. But the lungs of a sperm whale and other deep-diving marine mammals are more flexible, scrunching up without rupturing the blood vessels.

Another adaptation is the whale's ability to store a lot of oxygen in its blood and muscles. A protein in the blood called hemoglobin carries oxygen in the circulatory system, picking up oxygen in the lungs and delivering it to the body's cells and tissues. In muscles, a protein called myoglobin stores oxygen for later use. The blood of sperm whales contains a concentration of hemoglobin than is several times higher than most land animals, and sperm whale muscles have nearly 10 times higher concentrations of myoglobin. This stored, dissolved oxygen lets the whale hold its breath for remarkably long periods, permitting deep dives.

But sperm whales may not escape all the harmful effects associated with deep diving. The bends, also known as decompression sickness, occurs when a diver ascends too rapidly. As the diver rises toward the surface, the pressure drops, and if the pressure drops too quickly, dissolved gases in the diver's blood and tissue form bubbles—the sudden release of pressure causes these gases to come out of solution, similar to the fizz of carbon dioxide when the top of a pressurized, carbonated beverage is removed. The bubbles can create little pits in the bones. Michael J. Moore and Greg A. Early, researchers at WHOI, recently studied the bones of whales and found pits and erosions, telltale signs of decompression damage. Although Moore and Early cannot prove the damage came from deep diving, there is no evidence for any other disease. Writing in a paper published in a 2004 issue of *Science* titled "Cumulative Sperm Whale Bone Damage and the Bends," the researchers noted, "It therefore appears that sperm whales may be neither anatomically nor physiologically immune to the effects of deep diving."

During most of these dives the sperm whales are seeking prey such as relatively small squid, but as the circular scars on some whales attest, the animals do not shy from battling larger beasts. One of the largest of the deep is the giant squid.

Giant squid are, like sperm whales, animals that can abide both shallow and deep water, although giant squid live in the deep and only occasionally visit the surface. Squid are cephalopods that have a mantle—the body—and ten arms, two of which are longer than the others. Most species of squid are small, with a mantle length under 12 inches (30 cm). But giant squid can reach an enormous size, as indicated by the carcass that washed up in Newfoundland in 1878. Giant squid are carnivorous; they use suction cups on their arms, particularly the long pair, to catch prey and bring it to their mouth.

No one had ever seen a live giant squid in its native environment until recently. Japanese scientists Tsunemi Kubodera and Kyoichi Mori finally succeeded in taking some digital pictures of a giant squid taking bait the researchers had placed at a depth of 2,950 feet (900 m) off the Ogasawara Islands, south of Tokyo, in the Pacific Ocean. Kubodera and Mori also recovered part of a tentacle from the animal, and noted the creature "appears to be a much more active predator than previously suspected," as the researchers wrote in their report, "First-Ever Observations of a Live Giant Squid in the Wild," published in 2005 in the *Proceedings of the Royal Society B: Biological Sciences.* In 2006, Kubodera and his colleagues managed to videotape a giant squid attacking bait left at the surface.

In 2007, researchers reported marine observations that they acquired when they managed to tag both sperm whales and a species of squid, called Humboldt or jumbo squid, in the same area of the Gulf of California at the same time. Randall W. Davis of Texas A&M University and his colleagues tagged five sperm whales and three Humboldt squid. The results showed that sperm whales spend a lot of time in the same depths as the squid. During the day, Humboldt squid stayed in the range of 650–1,300 feet (200–400 m) the majority of the time, a depth range in which sperm whales also spent much of their time during their dives. At night, squid visited shallower waters more often but also spent about half their time in the depth range given above, during which time they seemed to be resting. Whales continued to dive in that range at night, with about 75 percent of their dives reaching these depths. In their report, "Diving Behavior of Sperm Whales in Relation to Behavior of a Major Prey Species, the Jumbo Squid, in the Gulf of California, Mexico," published in a 2007 issue of *Marine Ecology Progress Series,* the researchers wrote, "Diving behavior by whales is thus consistent with the idea that they feed on jumbo squid at depth during the day, and we suggest that deep nighttime foraging may target squid that are recovering from stress after recent surface activity and are therefore more susceptible to predation."

APPLICATIONS OF THE BIOLOGY OF EXTREME ENVIRONMENTS

The study of giant squid, sperm whales, and other organisms that live or venture in the ocean depths is just beginning. Marine scientists are gaining important knowledge about marine ecosystems, which will enhance the ability of the seafood industry to harvest the resources of the sea in a sustainable way. But the biology of extreme environments such as the deep sea has other important applications.

Bioluminescence has been incorporated into many research projects of biology. Molecular biologists, for example, have the ability to transfer a gene of DNA from one organism to another; such transgenic techniques have a variety of benefits, such as helping crops to resist insect damage by inserting a gene whose protein product wards off these insects. Researchers also insert foreign genes into organisms to study the functions and properties of that gene, or to make a lot of the protein that is encoded by the gene in order to study it or to employ it in medical procedures.

Sometimes the methods to transfer genes have a low success rate. For instance, the new DNA may not be taken up by many cells, so few of them actually incorporate the gene. An efficient method to determine which cells have succeeded is necessary so that researchers can separate these cells and "grow" them, letting them divide and produce a large quantity of daughter cells. One way to do this is to include a "reporter" gene along with the gene of interest. The reporter gene makes a protein that is easily observed, so researchers can quickly select the cells in which the DNA transfer has succeeded. Often the reporter will be a bioluminescence gene, so that given the right conditions—the presence of luciferin, oxygen, and ATP—cells that have incorporated the new DNA will emit light. It does not get much easier to spot a successful transfer than that!

Another important application involves assays. An assay to determine how often a gene is activated may also employ bioluminescence. In this case, researchers ensure that the bioluminescence gene is activated at the same time as the gene of interest. The amount of emitted light will therefore be related to the extent to which the genes are activated—a bright cell is much more active than a dark one.

Many of these techniques use a firefly gene that codes for luciferase, but researchers also make use of luciferase genes from marine organisms. Sometimes fluorescent proteins from jellyfish and other marine creatures are used instead. These genes and proteins have slightly different properties, and the experimental conditions and the nature of the cells and their milieu—their environment—usually dictate which genes and proteins should be used in a given procedure.

Sometimes none of the available genes are very effective. New techniques are also being developed in which bioluminescent imaging plays a role in evaluating the promise of such techniques as stem cell therapy, and improved bioluminescent materials are needed. The best place to look for these new materials is the deep sea, where so many of the organisms are bioluminescent.

No one can predict what kind of interesting and useful molecules are yet to be discovered. One of the most useful molecules ever found is Taq polymerase, an enzyme complex that is crucial for a technique known as polymerase chain reaction (PCR) that is critical in DNA forensics—for instance, the identification of criminal suspects based on DNA evidence—as well as genetic testing, genetic research, and many other important activities. Taq was named after *Thermus aquaticus,* a bacterium that lives in hot springs. Similar bacteria live around hydrothermal vents. This enzyme is able to withstand high temperatures—it lives in hot water, which accounts for its name (*thermē* is a Greek term meaning heat)—which is essential for PCR. In 1989, *Science* named this polymerase its "molecule of the year."

Organisms that live in the extreme environments of the deep sea have yet to be studied thoroughly, for reasons discussed above, and many species are still being discovered. Marine biologists are eager to explore and develop improved experimental procedures that can potentially uncover many more future molecules of the year.

CONCLUSION

Further exploration of deep-sea habitats will undoubtedly enrich scientific techniques in a variety of disciplines, especially genetics. Deep-sea research has already revealed an array of interesting and unusual features, such as bioluminescent animals that generate light to lure prey or find mates in the cold, dark abysses of the ocean.

Molecular biology and genetics can also help deep-sea research. Researchers have developed the capacity to rapidly sequence large segments of DNA, such as the entire set of DNA—the genome—of various organisms, including a human being. (Scientists working on the Human Genome Project finished sequencing the human genome in 2003.) Knowing the genome sequence is a great help in the study of an organism, particularly its genetic processes, but important lessons can also be learned by comparing genomes of different organisms. Comparing genomes helps scientists understand the biological mechanisms that organisms have in common as well as those that are special adaptations, unique to a particular organism or a group of organisms inhabiting an unusual environment.

Deep-sea organisms must thrive in dark environments of enormous pressure and generally cold temperatures. In 2005, Alessandro Vezzi and Giorgio Valle of the University of Padova in Italy, Douglas H. Bartlett of Scripps Institution of Oceanography, and their colleagues reported the genome sequence of a bacterium, *Photobacterium profundum,* which had been found at a depth of 8,200 feet (2,500 m). At this depth, an organism is subject to about 3,700 pounds (1,680 kg) per square inch (6.25 cm^2). To maintain a supply of these organisms for research purposes, the scientists grow them in high-pressure cylinders in the laboratory.

Sequencing a genome is a tremendous effort, but analyzing and understanding a genome is an even greater task. Biologists have made progress identifying genes and gene function but the process takes a considerable amount of time due to the number of genes and the complexity of their interactions. The size of the genome of *P. profundum* is about 0.2 percent that of the human genome, but any organism, even a bacterium, is complex. Vezzi, Valle, Bartlett, and their colleagues have found new genes associated with membrane proteins, particularly those involved in transport, which fits well with the findings discussed in the section titled "Adapting to the Depths." A number of other new genes with unknown functions have also been found. The researchers published their report, "Life at Depth: *Photobacterium profundum* Genome Sequence and Expression Analysis," in a 2005 issue of *Science.*

Biologists, chemists, oceanographers, and adventurers who love exploring the deep sea will continue to advance scientific knowledge as they study marine life in the ocean and search for new species. The ocean's great depths are hidden, yet the secrets are slowly yielding to science. The deep sea and its denizens are exciting for their scientific benefits as well as their allure as one of Earth's last remaining frontiers.

CHRONOLOGY

1818 C.E.   British explorer Sir John Ross (1777–1856) finds life in mud samples he retrieved from the ocean floor as deep as one mile (1.6 km) in Baffin Bay.

1830s   Because sunlight does not penetrate very far into the ocean, British scientist Edward Forbes (1815–54) develops the notion that the deep sea is generally an azoic zone, without life.

1872–76   The expedition of HMS *Challenger,* in which researchers circled the globe and collected many specimens, confirms the existence of organisms living in the deep sea.

| 1878 | A giant squid having a tentacle measuring at least 35 feet (10.7 m) washes ashore at Thimble Tickle Bay in Newfoundland. |

1930   Woods Hole Oceanographic Institution, which has conducted many important deep-sea investigations, is founded at Woods Hole, Massachusetts.

1977   A research team of French and American scientists, led by Robert Ballard, discover the first hydrothermal vent, located on the East Pacific Rise.

1987   David Packard (1912–96) establishes the not-for-profit Monterey Bay Aquarium Research Institute at Moss Landing, California. The institute has taken full advantage of its proximity to the deep submarine canyon located in Monterey Bay to explore deep-sea habitats.

2004   Japanese scientists Tsunemi Kubodera and Kyoichi Mori take the first pictures of a giant squid in its natural environment. The findings are published in 2005.

2006   Kubodera and his colleagues take the first video of a giant squid as it attacks bait left at the ocean surface.

2007   Randall W. Davis and a team of researchers report observations acquired when the scientists manage to tag both sperm whales and a species of large squid, called Humboldt or jumbo squid, in the same area of the Gulf of California at the same time.

FURTHER RESOURCES

Print and Internet

Brandt, Angelika, Andrew J. Gooday, Simone N. Brandão, Saskia Brix, Wiebke Brökeland, Tomas Cedhagen, et al. "First Insights into the

Biodiversity and Biogeography of the Southern Ocean Deep Sea." *Nature* 447 (May 17, 2007): 307–311. Scientists collected specimens from a depth range of 2,450–20,820 feet (748–6,348 m) and found hundreds of new species.

Brindley, Amanda A., Richard W. Pickersgill, Julian C. Partridge, David J. Dunstan, David M. Hunt, and Martin J. Warren. "Enzyme Sequence and Its Relationship to Hyperbaric Stability of Artificial and Natural Fish Lactate Dehydrogenases." *PLoS One* April 2008. Available online. URL: http://www.plosone.org/article/info%3Adoi %2F10.1371%2Fjournal.pone.0002042. Accessed June 9, 2009. The researchers report on their research on hyperbaric proteins.

Davis, R. W., N. Jaquet, D. Gendron, U. Markaida, G. Bazzino, and W. Gilly. "Diving Behavior of Sperm Whales in Relation to Behavior of a Major Prey Species, the Jumbo Squid, in the Gulf of California, Mexico." *Marine Ecology Progress Series* 333 (2007): 291–302. Researchers tag both sperm whales and jumbo squid to study their movements.

DeLong, Edward F., and A. Aristides Yayanos. "Adaptation of the Membrane Lipids of a Deep-Sea Bacterium to Changes in Hydrostatic Pressure." *Science* 228 (May 31, 1985): 1,101–1,103. The scientists report on adaptations in certain molecules of the membranes of deep-sea organisms.

Distel, Daniel L., Amy R. Baco, Ellie Chuang, Wendy Morrill, Colleen Cavanaugh, and Craig R. Smith. "Do Mussels Take Wooden Steps to Deep-Sea Vents?" *Nature* 403 (February 17, 2000): 725–726. Distel and his colleagues propose that certain organisms spread across the ocean floor from vent to vent by taking small "steps," which consist of whale falls and sunken wood.

Ellis, Richard. *The Search for the Giant Squid: The Biology and Mythology of the World's Most Elusive Sea Creature.* New York: Penguin, 1999. Although written before the first undersea video of a giant squid was taken, this book offers a richly detailed account of the observations and speculations of people who have searched for this mysterious beast over the past few centuries.

Haddock, Steven H. D., Casey W. Dunn, Philip R. Pugh, and Christine E. Schnitzler. "Bioluminescent and Red-Fluorescent Lures in a Deep-Sea Siphonophore." *Science* 309 (July 8, 2005): 263. In this single-page article, researchers describe siphonophores that wave and flick luminous filaments amid stinging tentacles.

Haddock, Steven, C. M. McDougall, and James F. Case. "The Bioluminescence Web Page." Available online. URL: http://www.lifesci.ucsb.edu/~biolum/. Accessed June 9, 2009. Haddock, a researcher at the Monterey Bay Aquarium Research Institute, and his colleagues developed this Web resource devoted to bioluminescence. The site discusses bioluminescent myths, organisms, chemistry, and physiology, and includes many illustrations and images.

Hoyt, Erich. *Creatures of the Deep: In Search of the Sea's "Monsters" and the World They Live in.* Buffalo, N.Y.: Firefly Books, 2001. Although Kraken and other legendary sea monsters that could take down a man-of-war are in short supply, there are plenty of fanged and ferocious entries in this well-illustrated, large-format volume.

Kubodera, Tsunemi, and Kyoichi Mori. "First-Ever Observations of a Live Giant Squid in the Wild." *Proceedings of the Royal Society B: Biological Sciences* 272 (2005): 2,583–2,586. The researchers describe the first sighting of a live giant squid.

MarineBio. "The Deep Sea." Available online. URL: http://www.marinebio.com/Oceans/TheDeep/. Accessed June 9, 2009. MarineBio, a nonprofit conservation and education group, maintains this informative Web page that describes deep-sea animals and their environments.

Monterey Bay Aquarium. "Exploring the Ocean with the Monterey Bay Aquarium Research Institute." Available online. URL: http://www.mbayaq.org/efc/efc_mbari/mbari_home.asp. Accessed June 9, 2009. Based on an exhibit at Monterey Bay Aquarium and expeditions of the Monterey Bay Aquarium Research Institute, this Web resource offers three virtual deep-sea explorations: discovering new species, mapping undersea mountains, and investigating the remains of a sunken whale.

Moore, Michael J., and Greg A. Early. "Cumulative Sperm Whale Bone Damage and the Bends." *Science* 306 (December 24, 2004): 2,215. In this single-page article, the researchers describe injuries of whale bone that appear to be caused by the effects of deep diving.

National Oceanic and Atmospheric Administration. "Sperm Whales." Available online. URL: http://www.nmfs.noaa.gov/pr/species/mammals/

cetaceans/spermwhale.htm. Accessed June 9, 2009. The wealth of information on this Web page describes the appearance, habitats, and behavior of these amazing animals.

Nouvian, Claire, ed. *The Deep: The Extraordinary Creatures of the Abyss.* Chicago: University of Chicago Press, 2007. This large-format book contains short essays on the remarkable organisms that live in the deep ocean, along with many color photographs that are, as the old saying goes, worth 1,000 words (or more).

Smith, Craig R., and Amy R. Baco. "Ecology of Whale Falls at the Deep-Sea Floor." *Oceanography and Marine Biology: An Annual Review* 41 (2003): 311–354. The researchers describe ecosystems that develop in and around whale carcasses.

Vezzi, A., S. Campanaro, M. D'Angelo, F. Simonato, N. Vitulo, F. M. Lauro, et al. "Life at Depth: *Photobacterium profundum* Genome Sequence and Expression Analysis." *Science* 307 (March 4, 2005): 1,459–1,461. The researchers report on the set of genes of a deep-sea organism.

Yancey, Paul H. "Deep-Sea Biology." Available online. URL: http://people.whitman.edu/~yancey/deepsea.html. Accessed June 9, 2009. Yancey, a professor at Whitman College in Walla Walla, Washington, has amassed a tremendous amount of information at this informative Web resource. Topics include research ships, deep-sea species, and how life has adapted to the high pressure of the ocean depths.

Web Sites

Monterey Bay Aquarium Research Institute. Available online. URL: http://www.mbari.org/. Accessed June 9, 2009. The Web site for MBARI contains news and information on its research, observatories, expeditions, and current projects, with plenty of images and maps.

National Museum of Natural History: In Search of Giant Squid. Available online. URL: http://seawifs.gsfc.nasa.gov/squid.html. Accessed June 9, 2009. This material is based on an exhibit presented at the National Museum of Natural History, part of the Smithsonian Institution. The text and illustrations document the discovery of this elusive deep-sea creature and research on its physiology and behavior.

4

TSUNAMI: KILLER WAVES

On December 26, 2004, a powerful earthquake shifted a large section of the seafloor in the Indian Ocean upward about 33 feet (10 m). The magnitude of this earthquake measured 9.2 on the moment magnitude scale (which has values similar to the old Richter scale, which the moment magnitude scale has replaced), one of the strongest earthquakes in history. As with most earthquakes, it occurred near a tectonic plate boundary—a *fault,* where gigantic, slowly moving plates of Earth's crust meet. In this case, the earthquake's origin was about 18 miles (30 km) below the ocean floor near the Sunda Trench, at the site where the Indian Plate dives beneath the Eurasian Plate, off the coast of Sumatra, Indonesia.

The sudden uplifting of the seabed pushed against the column of water above it, creating a disturbance that traveled quickly in all directions. Such a disturbance is called a tsunami—a series of waves that can travel at the speed of a jet airplane. The term *tsunami* comes from the Japanese words *tsu,* meaning harbor, and *nami,* meaning wave. (In Japanese, the plural of this word is also tsunami, but many English speakers add an "s" to indicate more than one. Note that a single tsunami usually consists of numerous waves.) Fifteen minutes after the earthquake, the tsunami struck the coast of Sumatra, an island of Indonesia. A short time later the waves reached parts of the coast of India, Sri Lanka, Thailand, and elsewhere. In some places, the water surged up to 100 feet (33 m). Residents had little or no warning, and about 250,000 people perished in the disaster.

Workers in Sri Lanka begin the laborious process of cleaning up in the aftermath of the Indian Ocean tsunami of 2004. *(AP Photo/Vincent Thian)*

Following the tragedy, officials installed a warning system for the affected areas, and governments scrambled to evaluate and upgrade systems already in place in other parts of the world. The technology to provide warning, and possibly forecasts, of tsunamis would be greatly improved if scientists knew more about how these disturbances are generated. High-magnitude undersea earthquakes are relatively common, but not all such earthquakes produce a significant tsunami. Huge waves can also arise from other sources, such as landslides; for example, an earthquake caused a landslide that fell into Lituya Bay in Alaska on July 9, 1958, sending a wall of water that hit the opposite shore and reached an astounding height of 1,720 feet (524 m)—higher than the Empire State Building in New York! Researchers study fast, giant waves in order to learn more about the ocean as well to help guard against future disasters. This chapter discusses what has been learned about the formation and effects of tsunamis, and how scientists and engineers are putting this knowledge to work.

INTRODUCTION

A wave is a propagating disturbance. Waves usually propagate through some medium, or material, such as sound waves through air or water. The original disturbance, such as a thunderclap in air or a rock tossed into a pond, sets into motion some of the particles of the medium, which jostle their neighbors, and so on, as the disturbance spreads through the material. (A medium is not always required—electromagnetic waves can propagate through empty space.) What spreads in a wave is the disturbance, not the particles; although the particles of the medium move, their motion is a vibration occurring over a limited range of space. The disturbance gets passed from particle to particle as they bump or interact with one another, conducting the energy of the wave through the material without any one particle traveling the whole distance.

In waves known as longitudinal, the particles vibrate back and forth along the direction of the wave's motion. Sound waves, for example, are longitudinal; the initial disturbance pushes out against air molecules, which collide with their neighbors and vibrate in the same direction as the wave propagation. These vibrations produce pressure variations as

Ocean wave *(Mana Photo/Shutterstock)*

air molecules periodically compress and spread out; these variations, when they impinge upon ears, are perceived as sound. Particle motion in waves known as transverse waves occurs perpendicular to the wave's direction, so that the particles or regions of the medium vibrate up and down, at a right (90°) angle to the propagation. Such waves occur along a taut string, such as a guitar string, when it is plucked.

Waves at the surface of the ocean tend to be a combination of longitudinal and transverse motions. Most ocean waves are created when wind blows over the surface, creating ripples. Stronger winds blow against the ripples, lifting them and making the wave larger. The up-and-down movement of the wave results in a crest, or peak of the wave, and then a trough. But the water at any given point of the propagating wave actually moves in a circular, or orbital motion, rather than a strictly vertical movement: Water moves slightly forward, then up to the crest, then back a little, and then down to form the trough.

Ocean waves propagate throughout the sea, and if strong enough, continue well beyond the area in which the wind created the initial disturbance. Waves that continue outside the windy area of origin tend to be rounded and are known as swells. Such waves usually end up lapping gently along shorelines.

The *wavelength* of a wave is the distance from one crest to another. Another property of a wave is its frequency, which is the number of crests that pass a specific point per unit of time. A wave's speed is the rate at which it travels, and can be determined by the wavelength times its frequency. For example, if the wavelength is 5 feet (1.5 m) and its frequency is 3 per second—three crests pass by a given point each second—then its speed is $5 \times 3 = 15$ feet/sec (4.5 m/s).

Waves have another property known as interference. If two waves occupy the same space at the same time, they add. This means that when crests overlap, the resulting wave is the sum of the two crests, and the troughs merge into one deeper trough. When equal-sized crests and troughs meet, they cancel, leaving the surface undisturbed. Interference becomes especially important in confined spaces such as harbors, where waves commonly overlap and produce significant crests and troughs. In the open sea, scientists think that rare events known as rogue waves are also due at least in part to interference. Rogue waves can be huge; for example, during a hurricane in 1995 the RMS *Queen Elizabeth II*, a cruise ship, encountered a monstrous wave that reached 95 feet (29 m). In a 2002 interview with the

British Broadcasting Corporation (BBC), Captain Ronald Warwick said of the wave, "Out of the darkness came this great wall of water. I have never seen a wave as big as this in my whole life." But the ship received only minor damage, and no one was seriously hurt in the incident.

A tsunami is not a wind-generated wave, and as discussed in the following section, is hardly noticeable in deep water. But tsunamis share other properties with ocean waves. A tsunami is usually composed of a series of waves or undulations, with a trough sometimes arriving at the shore first. When the trough arrives first, the water along the coast eerily recedes, enticing unwary beach-goers to venture out from the shore. This is a serious mistake—a crest is forthcoming, and the water level will quickly rise.

National Oceanic and Atmospheric Administration (NOAA)

In July 1970, then-president Richard Nixon announced in the Reorganization Plans Numbers 3 and 4, "We face immediate and compelling needs for better protection of life and property from natural hazards. . . . We also face a compelling need for exploration and development leading to the intelligent use of our marine resources. We must understand the nature of these resources, and assure their development without either contaminating the marine environment or upsetting the balance." With those words, President Nixon and his administration proposed the creation of the National Oceanic and Atmospheric Administration (NOAA), usually pronounced "Noah." NOAA was formally established on October 3, 1970.

But the roots of NOAA run much deeper in American history. Although NOAA became a new agency in 1970 within the Department of Commerce, it was formed by a reorganization of much older agencies. Among the agencies incorporated into NOAA at the time of its establishment were the United States Coast and Geodetic Survey (created in 1807),

The back and forth movement of the water during a tsunami resembles tides, and tsunamis are sometimes called tidal waves. Tides consist of periodic risings and fallings of the average sea level—at high tide the water encroaches far up the shore, and at low tide it recedes to its minimum. Gravitational forces of the Sun and especially the Moon, along with the motion of Earth, cause tides. (The Moon, although much smaller and less massive than the Sun, is particularly important in tidal processes because of its proximity to Earth.)

A tsunami is unrelated to the tides, and has a much different cause. And although the rising and lowering of water due to a tsunami resembles tidal movements, the magnitude of the changes can be much

the Weather Bureau (1870), and the Bureau of Commercial Fisheries (1871). These agencies were the earliest United States government organizations dedicated to marine and weather issues.

The main office of NOAA is located at Silver Spring, Maryland, close to Washington, D.C., but the agency has laboratories and observatories in many different places. NOAA is involved in all aspects of marine, weather, and climate research, including issuing daily weather forecasts, monitoring the climate, managing fisheries and marine commerce, exploring the world's oceans, and tracking severe storms. Hurricane season, which runs from June 1 to November 30—when storms most frequently form in the warm waters of the Atlantic Ocean—is a particularly busy time for NOAA.

NOAA also monitors tsunamis and conducts research aimed at gaining a better understanding of these powerful waves. The NOAA Center for Tsunami Research, located in Seattle, Washington, at the Pacific Marine Environmental Laboratory (PMEL), studies tsunami propagation as well as flooding and other effects tsunamis have when they reach shore, and also plays a role in the design and development of improved tsunami warning sensors. This research is critical in enhancing the world's preparedness for future tsunami events.

greater, and the speed at which they occur in a tsunami is tremendously more rapid.

The Pacific Ocean is Earth's largest body of water, accounting for about half of Earth's ocean surface, yet it hosts more than its share of these dangerous waves. More than 80 percent of tsunamis occur in the Pacific Ocean, and these dangerous waves are common in Japan, which experiences more tsunamis than any other country (explaining the Japanese origin of the term *tsunami*). The reason is that this region is geologically active, with many earthquakes and volcanoes occurring around the Pacific Ring of Fire as Earth's tectonic plates collide and shift. This activity generates a lot of tsunamis.

The Indian Ocean tsunami of 2004 served as a horrific reminder of the power and threat of tsunamis. Monitoring these killer waves is essential. The main government agency in the United States that studies and monitors the oceans is the National Oceanic and Atmospheric Administration (NOAA). This important agency maintains a Center for Tsunami Research, but as described in the sidebar, NOAA participates in diverse activities relating to the oceans and the atmosphere.

Huge, devastating tsunamis such as the one in 2004 in the Indian Ocean are rare, but smaller tsunamis are much more common. A tsunami occurs every few years on average, though most inflict little or no damage. The worst tsunami to hit the United States occurred on April 1, 1946, when a strong earthquake rocked the Aleutian Trench off the coast of the Alaskan island of Unimak. The resulting tsunami toppled a two-story lighthouse on the island, killing five people. But the severest effects in terms of American casualties came when the tsunami reached the Hawaiian Islands, more than 1,860 miles (3,000 km) away. Although somewhat reduced at this distance, the tsunami swept into the city of Hilo, causing major damage and killing 159 people. It was Hawaii's worst natural disaster in terms of lives lost.

Tsunamis have long elicited wonder. They have occurred throughout recorded history and long before, of course. One of the earliest and most devastating tsunamis in ancient history took place circa 1630 B.C.E. after a volcanic eruption on Santorini, a Greek island in the Aegean Sea. The explosion created waves that flooded the surrounding region, possibly contributing to the demise of a then flourishing culture, the Minoans, who lived in the eastern Mediterranean region. Other tsunamis have also been noted in ancient times. The Greek historian Thucydides (ca. 460–400 B.C.E.) wrote *The History of the Peloponnesian War* during the late fifth century or early fourth century B.C.E., and made the following comment about the aftermath of a series of earthquakes during the war: "About the same time that these earthquakes were so common, the sea at Orobiae, in Euboea, retiring from the then line of coast, returned in a huge wave and invaded a great part of the town, and retreated leaving some of it still under water." This is an excellent description of a tsunami.

Thucydides speculated that earthquakes cause these waves, and he was basically correct. Tsunamis occur when violent disturbances such as earthquakes move massive amounts of water.

THE PHYSICS OF TSUNAMIS

The waves of a tsunami propagate energy as do all waves, but tsunamis differ from wind-generated waves in certain properties. Most ocean waves do not extend very deep into the water, since the motion is confined to the surface and a short distance below. The bottom or ocean floor does not affect these waves until they reach shallow water, such as the shoreline or a reef. In shallow water, the bottom interferes with the wave motion, slowing its speed and decreasing the wavelength as the crests start to get closer together. Although some of the wave's energy is lost in this process, the vertical height increases as the water "bounces" off the bottom. When the wave gets too high to support itself, it breaks, with the crest spilling forward.

A violent event that shakes the ocean floor and sets into motion a tsunami affects the entire column of water over this area. As a result, tsunamis have exceptionally long wavelengths, often about 120 miles (200 km), and always "feel" the bottom. Scientists have found that the speed of such a water wave, which is sometimes called a long wave, depends only on its depth and increases in proportion to the square root of the depth. For example, at a depth of 20,000 feet (6,100 m), a tsunami travels at a speed of about 545 miles/hour (870 km/hr)—faster than most passenger jets! Such a wave could travel from one side of the Pacific Ocean to another in less than a day.

In the deep ocean, a tsunami is but a fast-moving ripple, often with a height of less than 2 feet (0.6 m). The initial disturbance imparted an enormous quantity of energy to the overlying column of water, and this energy is efficiently transferred to neighboring regions, but when the column is tall—as it is in deep water—there is a lot of mass to move. As

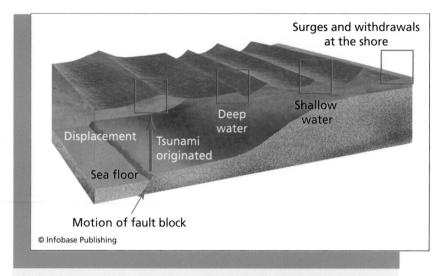

Surges and withdrawals
at the shore

Deep
water

Shallow
water

Displacement

Tsunami
originated

Sea floor

Motion of fault block

© Infobase Publishing

A disturbance, such as a sudden shift of a fault, creates the tsunami.
The waves propagate, rising in height as they reach the shore.

a consequence, the wave is not very high while traveling through deep portions of the ocean.

The height changes dramatically when the tsunami reaches shore. Shallow depths compress the wave and slow it down, similar to wind-generated waves, and some of the energy is lost. But a great deal of energy remains, and in shallow water the tsunami gains tremendous height and surges onto the coast as if it were an exceptionally high and fast tide. The figure illustrates tsunami generation and propagation. A tsunami can reach amazing heights; a 1971 tsunami in the Ryukyu Islands of Japan was 280 feet (85 m) high.

To trigger a tsunami, a large column of water must be displaced. This requires a sudden, vertical movement—up or down—of the seafloor, which occurs in certain types of earthquakes, volcanic eruptions, and undersea landslides (a mass of falling rock or mud under the surface of the ocean). Earthquakes tend to occur along plate boundaries, but not all undersea plate boundaries create the right conditions for tsunami-generating earthquakes. An earthquake caused by a sudden horizontal displacement, such as a sideways slippage along a fault, will not lift the overlying column of water, so it is less likely to generate a tsunami. (But if a tsunami does form, any lateral movement can certainly contribute energy to the process, increasing the size of the waves.) Most undersea plate boundaries that generate tsunamis are of the type known as sub-duction, where one plate is diving beneath another. Earthquakes tend to produce vertical displacements in these zones as one plate suddenly shifts downward. Boundaries where plates grind past one another tend to produce horizontal slippage and fewer tsunamis. Faults near Japan and Chile are subduction zones.

Studying the behavior of tsunamis is difficult because of the im-mense size and destructive power of these phenomena. Scientists and engineers have recently begun making miniature "tsunamis" in the laboratory, which permits greater observation and control over the variables. One of the largest tsunami laboratories in the world is the Tsunami Wave Basin at Oregon State University, which is 160 feet (50 m) in length, 87 feet (26.5 m) wide, and 7 feet (2.1 m) deep. An electric motor generates the waves.

Researchers from around the country visit the laboratory and con-duct experiments. For example, in 2007 Yin Lu Young and her students began a series of studies on the effects of waves on soil. These studies are important for many reasons, such as the effects on buildings and structures sitting on the soil, and their ability to withstand a tsunami.

Depiction of a giant tsunami striking a populated coastline
(Fouquin/Shutterstock)

Each experiment requires a substantial investment in time to set up the conditions and arrange the soil, so the work is slow and ongoing. Young and her colleagues, as well as other researchers who are conducting their own experiments, hope to acquire enough knowledge to prevent the collapse of buildings and infrastructure that commonly takes place in tsunami disasters.

LANDSLIDES AND METEORITES

An object falling into a body of water creates a splash. A large object, or a huge number of small ones, such as in a landslide, produces a phenomenon known as a splash wave. Splash waves share many properties with tsunamis, and many oceanographers classified them as a type of tsunami. The 1958 Lituya Bay wave, which swept up to a height of 1,720 feet (524 m) when it came ashore, is an example of a landslide-generated splash wave. This wave occurred when an earthquake dislodged a huge quantity of rock, some of which was as high as 3,000 feet (915 m), which fell into one end of the bay. Water sloshed to the other side of the bay, creating a towering wave—the highest ever recorded—witnessed by four people on two fishing boats in the bay. One of the boats rode the wave like a surfboard, while the other boat had its anchor line snapped and was driven onto land. A third boat, containing two people, was also in the bay at the time, but unfortunately capsized, killing the two people on board.

Special conditions at the bay, in addition to the landslide, probably contributed to the monstrous size of this wave. Lituya Bay is deep, narrow, and has a length of 7 miles (11 km). The landslide occurred at the back of the bay, and the wave raced along the narrow channel.

Material falling into the sea caused the Lituya Bay wave, but an underwater landslide can do the same thing. The ocean floor has abundant trenches, seamounts, and other rugged or uneven terrain. These features offer plenty of opportunities for landslides, especially during earthquakes, and many earthquake-generated tsunamis may actually be due, at least in part, to underwater landslides. (Difficulties in inspecting and surveying the depths contribute to the uncertainty in determining the precise nature of the event.) The continental shelf, as shown in the figure on page 6, seems particularly vulnerable.

University of Rhode Island researcher Stéphan Grilli and his colleagues are using mathematical equations, computer modeling, and laboratory experiments to study the properties of tsunamis that are created in underwater landslide events. Grilli employs a wave tank, located at the Ocean Engineering Department at the University of Rhode Island, which has a width of 12.1 feet (3.7 m), a depth of 5.9 feet (1.8 m), and a length of 98.4 feet (30 m). The "land" that slides is made of sheets of aluminum, formed into a smooth bell-shaped curve, and falls by gravity down a slope of about 15 degrees. During the experiments, Grilli and his colleagues measure landslide acceleration, water displacement, and the distance the water runs up on the "shoreline."

Researchers have derived mathematical formulas and computer models of tsunami generation based on physics and mechanics. Experimental configurations such as the Tsunami Wave Basin at Oregon State and Grilli's wave tank help scientists to test these ideas. Using sophisticated mathematics, Grilli and his colleagues have developed equations to predict tsunami wave properties, such as height and run-up, given the properties of underwater landslides, such as mass and distance. In a recent report, Grilli and his colleague François Enet of the Alkyon Hydraulic Consultancy and Research at Emmeloord, The Netherlands, found a good match between the predictions and the experimental measurements. The researchers published their report, "Experimental Study of Tsunami Generation by Three-Dimensional Rigid Underwater Landslides," in a 2007 issue of *Journal of Waterway, Port, Coastal, and Ocean Engineering*. Accurate models allow scientists to estimate the size and destructive power of waves that can be reasonably expected after landslides at various locations. The hazard can be quite serious, as discussed in the Conclusion section.

Another threat comes from the sky—the impact of bodies such as comets and meteorites falling into the ocean. Since about 70 percent of Earth's surface is ocean, astronomical bodies that hit the planet have an excellent chance of falling into water. Although no such impacts have affected Earth in recent times, the remnants of craters and other geological and fossil evidence clearly indicates the threat exists. Any low-lying coastal area is susceptible, but these events are exceedingly rare.

UNDERSEA EARTHQUAKES

Scientists believe that undersea earthquakes cause a lot of tsunamis because tsunamis often follow a major earthquake, as Thucydides noted about 2,400 years ago. But not all undersea earthquakes trigger a tsunami. Perhaps some or most of these earthquakes fail to generate a

The Indian Ocean Tsunami of 2004

The 9.2-magnitude earthquake that struck near Sumatra, Indonesia on December 26, 2004, could be expected to generate a serious tsunami. At the time of the disaster there was no warning system in place in the Indian Ocean, but scientists were monitoring the nearby Pacific Ocean, where the majority of tsunamis occur. Richard A. Kerr, a journalist writing in *Science* on January 14, 2005, noted that, "At PTWC [Pacific Tsunami Warning Center at Ewa Beach, Hawaii], staffers calculating magnitude from the seismic data circulating worldwide at first thought December's quake looked like a fairly run-of-the-mill magnitude 8.0." An earthquake of this magnitude could make local waves that would run up any nearby shores, as did the Papua New Guinea event in 1998, but would die out quickly. "So that first bulletin," Kerr wrote, "sent to participating Pacific Rim countries that PTWC is mandated to alert, reported the 8.0 magnitude and the absence of any threat around the Pacific."

An earthquake's magnitude requires a lot of data and some intensive study to calculate. The old scale, the Richter scale, judged earthquake magnitude by the size of the seis-

tsunami because the violent motion is more lateral than vertical, but without direct observation it is difficult to be certain. On other occasions, a tsunami that follows an earthquake is much stronger than would be expected based on the magnitude of the event. One example of this was a tsunami that struck Papua New Guinea in the southern Pacific Ocean on July 17, 1998, after an earthquake of magnitude 7.0 near the coastline. Waves as high as 50 feet (15 m) washed over the shore, killing more than 2,000 people. Although a magnitude of 7.0 is a strong earthquake, most events associated with tsunamis are larger. Researchers suspect a landslide caused by the earthquake may be responsible.

mic waves, but the new scale, the moment magnitude scale, categorizes events based on their energy, which geologists feel is a more accurate portrayal of intensity (and destructive force). Seismic waves provide an effective estimate of the moment magnitudes, but scientists often like to combine these measurements with other data, such as observations obtained in the field (which is not usually possible in undersea earthquakes). What generally happens immediately after an earthquake is that seismologists will make a preliminary announcement of magnitude, then revise this figure later, as more data becomes available.

Underestimating the 2004 Indian Ocean earthquake was an honest mistake, but not without consequences. As more data poured in, seismologists quickly raised the estimated magnitude. A few hours later scientists knew they had a much larger event on their hands, but by this time the tsunami had traveled across the Bay of Bengal, roughly 1,000 miles (1,600 km), and struck India and Sri Lanka. The waves also swept across the Pacific, efficiently transporting their energy until reaching parts of Asia, such as Thailand. A better understanding of undersea earthquakes and tsunamis might have helped prevent some of the tragedy.

The difficulty in accessing undersea terrain, combined with the complicated geological processes occurring on the seafloor, particularly around plate boundaries, has frustrated tsunami researchers. Powerful earthquakes have enough energy to set into motion a major tsunami, but as the Papua New Guinea experience in 1998 shows, other factors may come into play. Confusion also arises because of the difficulty in gauging the size of an undersea earthquake until scientists have had some time to study the records of their instruments. Because of the speed of tsunamis, any delay in perceiving the significance of an event can be catastrophic. Such delays contributed to the Indian Ocean disaster in 2004, as discussed in the sidebar.

Another event that surprised scientists occurred early in the morning of May 4, 2006, when an undersea earthquake of magnitude 7.9 struck near Tonga, a group of islands in the South Pacific Ocean. The size of the earthquake warranted issuing a tsunami warning, especially considering what had happened at Papua New Guinea in 1998, but fortunately nothing but tiny waves that did no damage followed the event. A short time later, a team of scientists from the United States, Australia, Japan, and Tonga set up seismic stations around the islands to record and study the aftershocks—a series of small earthquakes that usually follow a major one. The data suggested that the main earthquake had been due to a tearing of one section of the plate rather than a down- or upshift, which probably explains why no major tsunami occurred.

Other scientists are studying the ocean floor around plate boundaries in order to understand the forces and processes involved. If scientists can identify the mechanisms in a given region, they can gain a better idea of which areas are the most vulnerable. To study these processes, some inventive researchers have turned to techniques more familiar in biology—using sound to create images.

Ultrasound consists of sound waves with frequencies beyond the upper range of human hearing, which in young people tops out around 20,000 hertz (cycles per second). Physicians use ultrasound because high-frequency sound waves penetrate the soft tissues of the body, and reflections from the inner organs and tissues provides an image of the interior of the body, allowing doctors to "see" inside without performing an operation. These ultrasound procedures typically use sound waves with a frequency of a few million hertz and provide three-dimensional images. Checking the health of a fetus in pregnant women is one of the many uses of medical ultrasound.

Sound waves can also be used to image other structures. Gregory F. Moore, a University of Hawaii researcher who works with the Japan Marine Science and Technology Center, and his colleagues have recently analyzed three-dimensional images of the Nankai Trough, an underwater depression south of the Japanese island of Honshu, at the boundary where the Philippine Sea Plate is subducting (diving) beneath the Eurasian Plate. This region has been responsible for a number of tsunamis. To acquire the data, sailors aboard the M/V *Nordic Explorer* towed sound sources—in this case, airguns, which release pressurized air to create sounds of a broad range of frequencies—and hydrophones, which are instruments to record sound in water.

Moore and his colleagues used the images to study the cracks in the sediment layers and rocks. The researchers found visual evidence of a buried fault that scientists had long suspected because of the area's frequent earthquakes. Recent activity has moved upward, with slippage occurring all way up to or close to the ocean floor. Because of this increase in elevation, earthquakes and their associated movements have a bigger impact on the motion of the seabed. This motion is often vertical in subduction zones such as the Nankai Trough; as a result, previous tsunamis, such as one that occurred in 1944 that killed more than 1,200 people, have been severe. Future tsunamis can also be anticipated in this area. The researchers published their report, "Three-Dimensional Splay Fault Geometry and Implications for Tsunami Generation," in a 2007 issue of *Science.*

TSUNAMI FORECASTING

Knowledge gained in tsunami research is instrumental in helping scientists understand these spectacular waves. This research is also crucial in developing more effective preparations, in order to prevent or mitigate future disasters.

Forecasting a tsunami in a specific place at a specific time is not yet possible. Many scientists are skeptical that any such forecasting technique will soon be developed, because in most cases this would require the ability to predict the cause of most tsunamis—violent events such as undersea earthquakes—in advance. Although geologists commonly issue earthquake forecasts, the present state of Earth science does not allow for any greater precision than a certain probability. For instance, the United States Geological Survey, the main government agency devoted to Earth science, predicted in 2008 that California has better than a 99 percent chance of experiencing an earthquake of magnitude 6.7 or higher in the next 30 years. The prediction does not specify when in the next three decades the earthquake may occur.

Although the development of precise forecasts is unlikely in the near future, scientific research on tsunami generation and propagation will improve the ability to monitor the oceans and issue specific warnings. If people in the path of an approaching tsunami have advance warning, they can evacuate low-lying coastal areas. Even a short amount of time would be sufficient to save many lives.

The earliest major effort to detect tsunamis followed the 1946 event that struck Hawaii and Alaska, and resulted in the establishment of the

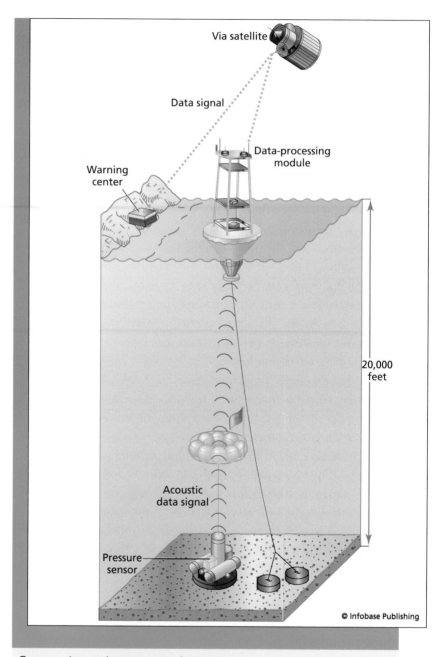

Sensors detect the pressure of a passing tsunami and relay the information to monitoring stations.

Tsunami Warning Center in 1949 at Ewa Beach, Hawaii, in the same facility as the Honolulu Geomagnetic Observatory. In 1960, after a severe tsunami struck Chile, many countries banded together, determined to reduce tsunami hazards. The United Nations and the Intergovernmental Oceanographic Commission founded the Intergovernmental Coordination Group for the Pacific Tsunami Warning Center in 1968, and the United States donated its tsunami warning facility at Ewa Beach as headquarters. The facility became known as the Pacific Tsunami Warning Center (PTWC). In 1967, the United States established the West Coast and Alaska Tsunami Warning Center, which assumed responsibility for these areas. The Indian Ocean lacked a warning system at the time of the 2004 disaster, but shortly afterward, officials installed a system in this region, which is currently the responsibility of PTWC.

Warning centers keep a close eye on seismic activity in their area of responsibility. If a significant earthquake occurs that could potentially cause a tsunami, officials issue a tsunami watch. The issuing of a watch will alert people to the possibility that a tsunami may have been generated, although its existence is not yet confirmed.

To determine if a tsunami has formed, PTWC relies on a network of advanced sensors. These sensors are generally below the water surface—remember that a tsunami traveling in deep water does not rise very far above the surface, and can pass a ship without the crew being aware of it. But the passing wave creates a pulse of pressure, which triggers extremely sensitive instruments on the ocean floor. As illustrated in the figure, a deep-sea sensor uses sound waves to contact a surface buoy, which in turn uses electromagnetic waves to transmit the information to a satellite, and from there on to the computers installed at the warning center. (Recall that most electromagnetic waves such as radio do not propagate very well in water, which is why the sensors transmit with sound.) This network of sensors is part of a program known as Deep-ocean Assessment and Reporting of Tsunamis (DART), which was developed by PMEL. Upon receiving evidence of a tsunami, the warning center issues its highest alert message, a tsunami warning.

Y. Tony Song of the Jet Propulsion Laboratory in Pasadena, California has recently devised a method based on global positioning system (GPS) stations that could improve tsunami warnings. A GPS receiver

Tsunami buoy ready for deployment on NOAA vessel *(AP Photo/NOAA)*

uses the transmissions of a set of satellites to determine its exact position on Earth. Analyzing data obtained from coastal GPS stations, Song computed the sea floor displacements created in recent tsunami-generating earthquakes, as reported in his research paper, "Detecting Tsunami Genesis and Scales Directly from Coastal GPS Stations," published in a 2007 issue of *Geophysical Research Letters.* By continuously monitoring this data, Song argues that future tsunamis can be observed shortly after generation. As Song notes in the report, "Because many GPS stations are already in operation for measuring ground motions in real time as often as once every few seconds, this study suggests a practical way of identifying tsunamigenic [tsunami-generating] earthquakes for early warnings and reducing false alarms."

Researchers at the University of Ulster in the United Kingdom, *Istituto Nazionale di Geofisica e Vulcanologia* [National Institute of Geophysics and Volcanology] in Rome, Italy, and California Institute of Technology have also studied seismic movements in relation to tsunami generation. The scientists, led by John McCloskey at the University of Ulster, used a sophisticated model to simulate ocean floor displacements and tsunami wave heights in about 100 scenarios that may occur in the vicinity of Sumatra, Indonesia, the same area as the 2004 Indian Ocean tsunami. The model incorporated the geological features of the Sunda Trench in order to conduct realistic simulations.

McCloskey and his colleagues found that for the region in the vicinity of the earthquake—the "near-field"—the shape of the waves is not strongly correlated with the location of the slippage along the fault or the magnitude of the earthquake. The tsunami wave height, however, is proportional to the amount of vertical movement. This work suggests that vertical displacement is key to an estimation of the strength of a tsunami in the Sunda area.

Other projects include a Short-term Inundation Forecasting for Tsunamis (SIFT) system, a research effort of PMEL scientists. The idea is to increase the accuracy of forecasts of tsunami height and coastal inundation (flooding) by real-time analysis of information from sources such as DART, the network of deep-sea sensors. Real-time analysis refers to an analysis conducted as the data comes in, the results of which are displayed in "real time" as opposed to a later, post-event analysis and display.

SIFT researchers hope to design formulas to estimate the time of arrival of the waves, their approximate height, and how far they will invade the shore. To help model a tsunami, researchers rely on data collected from previous tsunamis, assembled into a database for easy access and searching. For the inundation calculations, forecasters need to determine which areas will be hit, and then retrieve information on the properties of the targeted shores, such as the shape of the coastline and the average depth near the beach. This data, combined with the estimated wave height, would guide tsunami-warning officials in their evacuation effort.

The work of these and other researchers is aimed at improving the accuracy, speed, and reliability of tsunami forecasts and warnings. A better understanding of how these powerful waves are generated and propagate onto low-lying coastal regions is essential to gauge timely reactions to the threat, and eliminate unnecessary alarms or evacuations when there is no danger.

CONCLUSION

Although coastal regions in the vicinity of earthquake-prone faults experience the most trouble with tsunamis, earthquakes and other violent events can happen anywhere. Under certain conditions, these events can generate waves of astonishing height, such as the 1,720-foot (524-m) wave in Lituya Bay, Alaska, in 1958. The more scientists discover about how these waves form, the better they can identify areas that are at special risk.

Gary M. McMurtry at the University of Hawaii and his colleagues have provided some evidence that Hawaii has experienced waves as high as the 1958 Lituya Bay incident. McMurtry and his coworkers examined marine fossils found on the Kohala, the oldest volcano on the island of Hawaii (the "Big Island"). Using radioisotope dating—a technique that estimates the age of a substance by measuring the amounts of radioactive materials—the researchers determined that these fossils were about 110,000 years old, plus or minus 10,000 years. This part of Kohala has been slowly subsiding for some time, and in that long-ago era, the region where these fossils were found was more than 1,310 feet (400 m) above sea level and 3.7 miles (6 km) inland.

How did these relatively recent (in terms of geology) marine organisms reach such an elevated section of land? McMurtry and his colleagues believe that a tsunami is the most likely explanation. They point to a major landslide known to have occurred around this time—sonar mapping has revealed that a volume of earth measuring about 83 miles³ (350 km³) slid off the flank of Mauna Loa, a nearby volcano, around 120,000 years ago. A tsunami generated in this landslide could have sent water high up the coast, depositing the marine organisms whose fossils McMurtry and his colleagues studied.

Future tsunamis in the area are possible. The Hawaiian Islands are volcanic, and have some of the tallest mountains in the world. Mauna Kea, a dormant volcano, rises 13,796 feet (4,206 m) above sea level. But the island also extends about 19,500 feet (5,945 m) above the ocean floor, and if one includes this distance, Mauna Kea rises to a total height of 33,296 feet (10,151 m), which would be the tallest structure on Earth! (The summit of Mount Everest, the tallest mountain on land, is 29,028 feet [8,850 m] above sea level.) Volcanoes contain magma channels and chambers that can become hollow as the volcano ages, leading to a collapse and subsequent landslides. An underwater landslide on the flanks of such a massive structure could result in a giant tsunami, sending water sweeping over large portions of the island, and possibly reaching the west coast of the continental United States.

Another potential threat exists on the other coast of the United States. The east coast borders the Atlantic Ocean, a body of water with few tsunamis compared to the Pacific Ocean since the plate boundaries running down the middle of the Atlantic Ocean are diverging—the plates are separating—which is not favorable for tsunami-generation. But Steven N. Ward at the University of California, Santa Cruz, and Simon Day at the University College, London, in the United Kingdom, have identified a situation that merits attention. At the heart of the matter is Cumbre Vieja, a volcanic ridge on the small Atlantic Ocean island of La Palma in the Canary Islands, off the coast of Africa. (The name of the volcano is Spanish, meaning "old summit.")

Cumbre Vieja rises about 6,600 feet (2,000 m) above the Atlantic Ocean with a slope of about 15–20 degrees on average. It is an active volcano, erupting in 1949 and 1971. Ward and Day hypothesize that a major eruption could lead to a collapse of the volcano's western flank, which faces toward the east coast of the American continents.

A large volume of material falling from the steep flank of Cumbre Vieja would drop into the Atlantic Ocean. Ward and Day estimate that a volume as great as 119 miles^3 (500 km^3) might fall into the sea. As with the Lituya Bay landslide in 1958, a huge splash wave would result. The researchers calculate that such a wave may have a height of up to 82 feet (25 m) when it reaches the American coastline. If this is the case, water would surge onto the shore and flood many parts of the United States.

Devastating events are rare, but do occur, and scientists have discovered numerous incidents in Earth's history that have caused widespread destruction, such as the extinction of the dinosaurs about 65 million years ago. Scientists and officials do not want to spread undue alarm over low-probability events, yet it is wise to exercise caution—a potential threat should not be completely ignored just because it is unlikely to happen or happens only rarely. In 2008, NOAA completed an upgrade of the tsunami-warning network of buoys that provides alerts to coastal regions of the United States along the Pacific and Atlantic Oceans, including the Caribbean Sea and the Gulf of Mexico.

Tsunamis are fascinating waves. They are the products of events that generate an enormous quantity of energy, which spreads through the sea at a remarkable speed. A tsunami inspires a great deal of awe—and poses an equal amount of danger. Civilizations have been suffering from tsunamis since ancient times, but the expanding frontiers of marine science are gradually minimizing the threat.

CHRONOLOGY

ca. 1630 B.C.E. An explosion on Santorini, a Greek island in the Aegean Sea, produces giant waves that wash over the surrounding region, possibly contributing to the demise of the Minoan civilization.

ca. 400 B.C.E. Greek historian Thucydides (ca. 460–400 B.C.E.) provides a clear description of a tsunami in *The History of the Peloponnesian War,* and speculates that these waves are associated with earthquakes.

1946 C.E. A strong earthquake occurs off the Alaskan coast, resulting in a tsunami that kills 165 people in Alaska and Hawaii.

1949 The United States establishes a tsunami warning system at Ewa Beach, Hawaii.

1958 The largest recorded wave in history sweeps through Lituya Bay in Alaska, reaching a height of 1,720 feet (524 m).

1968 The United Nations and the Intergovernmental Oceanographic Commission found the Pacific Tsunami Warning Center in 1968, which incorporated the United States facility at Ewa Beach, Hawaii.

1970 The United States government, in recognition of the importance of studying and maintaining the marine environment and ocean resources, creates the National Oceanic and Atmospheric Administration (NOAA).

2004	An undersea earthquake near Sumatra, Indonesia, generates a powerful tsunami that inundates parts of Indonesia, India, Sri Lanka, Thailand, and other countries, killing more than 250,000 people.
2006	The United Nations coordinates the installation of a tsunami warning system in the Indian Ocean.
2008	NOAA completes an upgrade of the sensor network that monitors the generation and propagation of tsunamis posing a threat to the United States.

FURTHER RESOURCES
Print and Internet

Blackhall, Susan. *Tsunami.* Surrey, U.K.: TAJ Books, 2005. Following a brief general discussion of tsunamis, this book offers an extensive account of the Indian Ocean tragedy in 2004.

British Broadcasting Corporation. "Freak Wave—Transcript." November 14, 2002. Available online. URL: http://www.bbc.co.uk/science/horizon/2002/freakwavetrans.shtml. Accessed June 9, 2009. This transcript of a BBC Two program explores the nature of "freak waves."

Dudley, Walter C., and Min Lee. *Tsunami!* 2nd ed. Honolulu: University of Hawaii Press, 1998. Although focusing on tsunamis in Hawaii, the book also examines events in Japan, Chile, Alaska, the Mediterranean Sea, and Papua New Guinea, including eyewitness accounts. Note that this edition does not include information on the 2004 Indian Ocean tsunami.

Enet, François, and Stéphan T. Grilli. "Experimental Study of Tsunami Generation by Three-Dimensional Rigid Underwater Landslides." *Journal of Waterway, Port, Coastal, and Ocean Engineering* 133 (2007): 442–454. The researchers develop and test equations to predict tsunami wave properties, such as height and run-up.

Kerr, Richard A. "Failure to Gauge the Quake Crippled the Warning Effort." *Science* 307 (January 14, 2005): 201. This single-page news item describes failures during the Indian Ocean tsunami of 2004 that probably elevated the death toll.

Mathez, Edmond A., ed. *Earth: Inside and Out*. New York: New Press, 2001. Written by a team of experts, this highly informative book contains sections on Earth's evolution, seismic exploration of the interior, plate tectonics, analysis of rocks, and climate change.

McDaris, John. "Tsunami Visualizations." Available online. URL: http://serc.carleton.edu/NAGTWorkshops/visualization/collections/tsunami.html. Accessed June 9, 2009. Hosted by Carleton College, this Web resource offers a number of animations and movies that let the viewer track the course of several recent tsunamis, including the 2004 Indian Ocean event.

Moore, G. F., N. L. Bangs, A. Taira, S. Kuramoto, E. Pangborn, and H. J. Tobin. "Three-Dimensional Splay Fault Geometry and Implications for Tsunami Generation." *Science* 318 (November 16, 2007): 1,128–1,131. The researchers report on visual evidence of a buried fault in the Nankai Trough.

Prager, Ellen J. *Furious Earth: The Science and Nature of Earthquakes, Volcanoes, and Tsunamis*. New York: McGraw Hill, 1999. This book offers an excellent introduction to the geological processes of violent, earth-shaking events, including tsunamis.

Reorganization Plans Numbers 3 and 4 of 1970: Message from the President of the United States. July 9, 1970. Available online. URL: http://www.epa.gov/ocir/leglibrary/pdf/created.pdf. Accessed June 9, 2009. President Nixon describes his plans to reorganize government, including the establishment of NOAA.

Song, Y. T. "Detecting Tsunami Genesis and Scales Directly from Coastal GPS Stations." *Geophysical Research Letters* 34 (2007). Available online. URL: http://www.agu.org/pubs/crossref/2007/2007GL031681.shtml. Accessed June 9, 2009. Song describes a method based on global positioning system (GPS) stations that could improve tsunami warnings.

Thucydides. *The History of the Peloponnesian War*. Available online. URL: http://classics.mit.edu/Thucydides/pelopwar.html. Accessed June 9, 2009. Written in ancient times by a participant, this account describes the war between the Greek city-states.

Web Sites

Center for Tsunami Research. Available online. URL: http://nctr.pmel. noaa.gov/. Accessed June 9, 2009. This NOAA research program conducts research and develops methods to improve warning systems and minimize tsunami hazards. The Web site includes information on the center's research in hazard assessment and tsunami forecasting.

National Oceanic and Atmospheric Administration: Tsunami. Available online. URL: http://www.tsunami.noaa.gov/. Accessed June 9, 2009. This Web site presents news and information on all aspects of these dangerous waves, including generation and propagation, warning systems, forecasts, tsunami research, and recent tsunamis.

Pacific Tsunami Museum. Available online. URL: http://www.tsunami. org/. Accessed June 9, 2009. The Pacific Tsunami Museum, located in Hawaii, exhibits artifacts from past Hawaiian tsunamis and presents scientific information on tsunamis and their aftermath. At the museum's Web site, the visitor will find stories of tsunami survivors and photographs of tsunami damage.

Pacific Tsunami Warning Center. Available online. URL: http://www. prh.noaa.gov/ptwc/. Accessed June 9, 2009. Part of the National Weather Service (a division of NOAA), the Pacific Tsunami Warning Center keeps a close eye on tsunamis in the Indian Ocean and Caribbean Sea, as well as the Pacific Ocean. Their Web site lists recent tsunami messages and warnings.

United States Geological Survey: Tsunami and Earthquake Research at the USGS. Available online. URL: http://walrus.wr.usgs.gov/ tsunami/. Accessed June 9, 2009. The United States Geological Survey is the main government agency responsible for studying geological formations, natural resources, and natural hazards of the environment. Earthquakes are a prominent topic for the agency, and earthquake-generated tsunamis are discussed on this Web site. Links point to the latest research news and a number of tsunami animations.

5

EL NIÑO AND WEATHER

A fishing expedition off the coast of Washington in September 1997 netted a highly unusual catch for these waters—a striped marlin. Anglers prize these large game fish, which are normally found in warm, tropical waters. The 1997 event was the first recorded catch of a striped marlin off the coast of Washington. This part of the Pacific Ocean, normally quite chilly, had briefly warmed up.

Other unusual weather patterns arose during 1997 and 1998. Temperatures in many parts of the eastern Pacific Ocean averaged 9˚F (5˚C) higher than normal. Droughts and wildfires struck Indonesia. Seal and fish populations off the coast of Peru fell sharply. In the United States, heavy rains drenched most of California, causing serious flooding and mudslides. But these and many other uncommon events did not come entirely as a surprise, for they seemed to be related to a phenomenon that scientists have been studying for decades.

Fishermen along the coast of Peru have long known that a current of unusually warm water invades their coastal waters on occasion. It happens irregularly, once every two to seven years. This current carries fewer nutrients, which causes the fish populations to plummet, along with carnivores such as seals that feed off them. But these periodic events are also associated with increased rainfall in normally arid regions of the country, dramatically increasing crop yields. Since the phenomenon tended to begin around Christmas, Peruvians called it El Niño—Spanish for little boy, which referred to the birth of Christ. The reverse situation, in which the temperature of coastal Peruvian waters falls below normal, also occasionally occurs, although people did not notice it until much later. This cooling phase is the opposite of El Niño, and is known as *La Niña,* a Spanish term meaning little girl.

El Niño clearly has a major impact on Peru. In the 20th century, scientists began to realize that these events have a much greater sphere of influence and are associated with activity in Earth's atmosphere as well as the sea. Due to interactions between air and water, and the intricate links between the weather of different parts of the world, a change in one place can lead to significant changes in other, quite distant places. This chapter discusses how researchers study these phenomena, the methods used to track episodes of El Niño and La Niña, and the mechanisms by which these events are linked with the world's weather patterns. Curt Suplee, writing in the March 1999 issue of *National Geographic,* noted that after the 1997–98 episode had run its course, "the giant El Niño of 1997–98 had deranged weather patterns around the world, killed an estimated 2,100 people, and caused at least 33 billion (U.S.) dollars in property damage." A better understanding of these phenomena would be an important advance in marine science, and lead to improved weather and climate forecasts.

INTRODUCTION

El Niño and its associated phenomena have a long history of affecting society, although early on, people did not realize it. North Carolina State University archaeologist Scott M. Fitzpatrick and Richard Callaghan, a researcher at the University of Calgary in Canada, have found historical evidence that El Niño influenced the first voyage around the world.

In 1519, Portuguese explorer Ferdinand Magellan (1480–1521) led an expedition of five ships whose goal was to sail from Spain to the Spice Islands. Magellan and his crew reached the Pacific Island of Guam in 1521, well north of their intended destination. Historians have long wondered why Magellan took such a lengthy detour, but Fitzpatrick and Callaghan may have found an answer. The researchers studied historical records and also simulated weather patterns on a computer to recreate the prevailing conditions. When Magellan rounded the southern tip of South America, he sailed into smooth seas, which Fitzpatrick and Callaghan believe was the result of an El Niño occurring at the time. Magellan stuck close to the South American coast as he sailed northward instead of heading out into the open sea, possibly because he wanted to take advantage of the calm weather.

Another factor in the voyage was that Magellan got news of food shortages in the Spice Islands, so he stopped in Guam to provision the ship. Fitzpatrick and Callaghan believe the shortages were probably true, since they noted that the effects of El Niño often include drought in these areas. Magellan was killed in the Philippines, and only one ship returned to Spain after continuing the voyage all the way around the globe. The expedition suffered a severe loss of sailors and ships, but the first circumnavigation of the world was an amazing achievement—influenced, according to Fitzpatrick and Callaghan, by a then unknown El Niño.

Over time, weather records and ocean measurements became more detailed. In the 19th century, *meteorology*—the study of the atmosphere and weather—grew in importance as many nations set up observatories. Marine scientists became interested in Peru's El Niño after a particularly strong episode in 1891, although at that early stage of research, people thought of it as a local phenomenon. Another major episode occurred in 1925.

By 1957, researchers had started to notice other unusual patterns of activity occurring along with El Niño. That year had a significant El Niño, but just as importantly, scientists were making global observations at the same time. The International Council of Scientific Unions, a nongovernmental organization of scientific associations and unions, designated the 18-month period from July 1, 1957, to December 31, 1958, as the International Geophysical Year. (The International Council of Scientific Unions is now called the International Council for Science.) This designation encouraged scientists from all over the world to coordinate their observations and activities. The International Geophysical Year generated a lot of scientific activity, which, by chance, occurred at the same time as a strong El Niño. In the article "El Niño and La Niña," which the National Academy of Sciences released in March 2000, authors Roberta Conlan and Robert Service wrote, "Among the data they gathered were not only atmospheric measurements but also sea surface temperatures throughout the Pacific. . . . Some researchers in the 1950s noted that high sea surface temperatures off the coast of Peru seemed to correlate with a small difference in pressure across the tropical Pacific."

Atmospheric pressure is the force per unit of area exerted by the air. Slight variations in atmospheric pressure, such as regions of high or low pressure, play an important role in weather systems such as storms. These observations gave researchers a broader perspective of El Niño. Correlations with other changes happening elsewhere in the world suggested El Niño encompassed a much larger scale than previously believed. One of these changes was related to a periodic oscillation in atmospheric pressure that had been noted and named in the 1920s.

SOUTHERN OSCILLATION—EARTH'S ATMOSPHERE AND EL NIÑO

Sir Gilbert Walker (1868–1958) was a British mathematician and meteorologist. In 1904, Walker became Director General of Observatories in India. One of Gilbert's tasks was to investigate the Indian Ocean monsoons, which are seasonal winds that bring heavy rainfall to the region, and are critical components of the climate. Walker pored over decades of meteorological data from all over the world, including temperature, rainfall, and atmospheric pressure. One of Walker's most important findings was a back-and-forth relationship between the atmospheric pressure in the South Pacific, near Tahiti, and the Indian Ocean, west of Darwin, Australia. If pressure was high in the South Pacific, it was low in the Indian Ocean, and vice versa. Walker called this seesaw effect the Southern Oscillation.

Measurements and observations in the 1950s, mentioned in the previous section, and continuing through the 1960s, began to relate this atmospheric phenomenon with El Niño and its opposite, La Niña. The work of Jacob Bjerknes (1897–1975), a meteorologist at the University of California Los Angeles (UCLA), was particularly important, and led to a significant advance in understanding El Niño. Bjerknes linked El Niño with Walker's Southern Oscillation.

One of the phenomena that intrigued Bjerknes was variation in rainfall, which had been one of Walker's primary interests as well. Careful study of weather records led Bjerknes to the conclusion that significant variations in rainfall in certain regions around the Pacific Ocean corresponded to El Niño episodes. From there, he traced a connection to wind patterns.

Wind plays many critical roles in Earth's geology, weather, and climate. Differences in air pressure create wind. The movement follows a "downhill" direction, from high pressure to low. Consider, for an example, an extreme case, such as a vacuum—an empty space, devoid of air. When the vacuum is exposed to the atmosphere, air rushes to fill it, creating a stiff wind. (Other small objects get sucked in as well, which is the principle behind the vacuum cleaner.) Slight differences in pressure also generate wind, though not as strongly.

What creates differences in Earth's atmospheric pressure in the first place? The main engine driving Earth's atmosphere is energy from the Sun, delivered in the form of sunlight. Rays of sunlight warm the ocean or the ground, heating the air above it. Hot air rises, then cools in the upper atmosphere, and sinks. As described in the following sidebar, air circulates in fairly predictable patterns across the globe.

Bjerknes noticed a wind pattern that spanned the Pacific Ocean. He called this pattern the Walker Circulation, in honor of Sir Gilbert Walker. As illustrated in part (a) of the figure, this pattern circulates air between the western and eastern parts of the Pacific. On one side, near Australia, warm water causes air to rise, leaving an area of low pressure. On the other side of the Pacific Ocean, near the coast of South America, the air descends over cooler water, creating an area of high pressure. Winds generally blow from east to west; these steady winds, often known as trade winds, have aided many sailors on westward voyages.

But sometimes the conditions change. In El Niño episodes, as shown in part (b) of the figure, the pressure in the western Pacific Ocean increases, decreasing the winds and sometimes even reversing them, and the warm water current develops. The reason why this happens is discussed below, in the section "Forecasting the Oscillations." During La Niña, the cool water in the eastern Pacific Ocean spreads farther than usual, as shown in part (c) of the figure. La Niña is a more extreme version of typical conditions.

With the broadened scope of his research, Bjerknes realized that the pressure changes, as noted by Walker, and the spread of warm or cool water off the coast of Peru and in nearby regions, are parts of the same cyclical system. This cycle is often called *El Niño/Southern Oscillation* (ENSO). At irregular intervals, El Niño conditions prevail, but then change, reverting to the more typical conditions depicted in part (a) of the figure. The cycle may then go far in the other direction, initiating a La Niña event. An El Niño episode typically lasts about a year, and La Niña a little longer.

Wind and Circulation Cells

Consider a sea breeze, often enjoyed by inhabitants of coastal regions. On a sunny day, a refreshingly cool breeze commonly blows from the sea to the shore. This flow of air is the result of a rising of warm air over land during a sunny day; ground gets hotter than a water surface because water has a higher heat capacity (it can absorb more heat with less of a temperature change). The warm air rises, which results in less air over land. The temperature of the air column decreases over the cooler ocean, and the air sinks, leading to a slightly greater pressure. A gentle sea breeze, flowing from sea to shore, equalizes the pressure. At night, the opposite situation occurs, and a slight wind blows from shore to sea.

Similar situations occur on a larger scale all over the planet. The sun's rays hit tropical zones around the equator straight on, or nearly so, delivering more energy per unit area than in regions at or near Earth's poles. Air over the tropics is warm, and it expands and rises, then descends over cooler areas. The rotation of the planet beneath these air masses imparts additional motion and direction, producing high-velocity winds in the upper atmosphere called jet streams. The spinning Earth, along with the winds, also drives currents such as the Gulf Stream, in which a warm flow of water from the Gulf of Mexico crosses the North Atlantic Ocean.

Trade winds are also important. These winds prevail around the equator and the tropics, and tend to blow from the east. Christopher Columbus (1451–1506) noted these steady breezes on his journey across the Atlantic. The captains of sailing vessels subsequently came to know these winds well, which helped them on their journeys to the west.

Air and water currents distribute heat across the planet, resulting in less temperature variation than would occur otherwise. On the planet Mercury, which has no atmosphere or ocean, the temperature of the side facing the Sun can rise to about 752°F (400°C), while the night side can fall to -300°F (-184°C)—quite a difference!

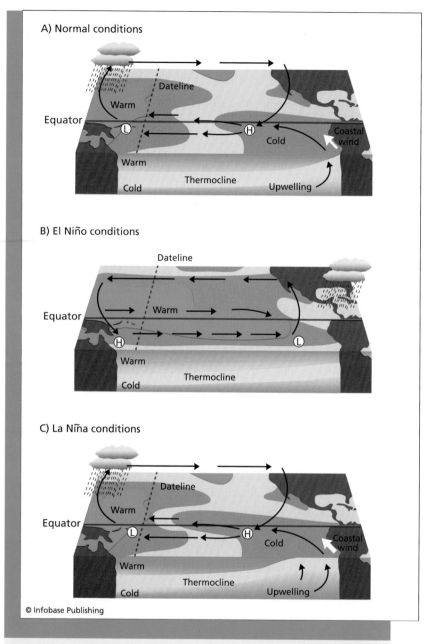

A) Normal conditions

Dateline

Warm

Equator

L

H

Cold

Coastal wind

Warm

Cold

Thermocline

Upwelling

B) El Niño conditions

Dateline

Equator

Warm

H

L

Warm

Cold

Thermocline

C) La Niña conditions

Dateline

Warm

Equator

L

H

Cold

Coastal wind

Warm

Cold

Thermocline

Upwelling

© Infobase Publishing

(a) Warm air rises in the eastern Pacific Ocean, leaving a zone of low pressure (L); the air cools and descends over the western portion of the ocean, an area of high pressure (H). (b) During an El Niño, the normal wind pattern slows or reverses; as a result, warm water invades coastal Peru, reducing the upwelling of cool, nutrient-rich water from below. (c) A La Niña episode intensifies the normal pattern, as cool water spreads beyond the South American coast.

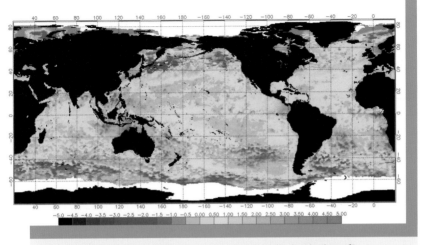

This image shows sea surface temperatures at the beginning of an El Niño in 2006. Note the warmer than usual water (denoted in red) off the northwest coast of South America. *(NOAA/Photo Researchers, Inc.)*

Why does this occur in the Pacific Ocean and not the Atlantic or Indian Oceans? Some types of oscillations do occur in other bodies of water, and researchers are not sure why El Niño phenomena are so pronounced in the Pacific Ocean. But the size of the Pacific Ocean, which is about twice as wide as the Atlantic, is probably important.

El Niño belongs to an extremely important pattern of interactions between the ocean and the atmosphere. In her 2002 book *El Niño: Unlocking the Secrets of the Master Weather-Maker,* J. Madeleine Nash wrote, "As scientists now understand it, El Niño is one side of a naturally occurring cycle that exercises the largest month-to-month influence on earthly weather patterns after the seasonal march of the earth about the sun."

Why did it take so long for people to discover this important cycle? One reason is that the cycle is so irregular—it does not occur in a regularly repeating pattern. The part of the cycle called El Niño may repeat in two years, or seven, or perhaps longer. Its intensity, as well as its occurrence, also varies; some episodes may be strong, as in 1982–83 and 1997–98, and others hardly noticeable. Another reason is that in the past, people were used to events of a local nature, not ones that spread

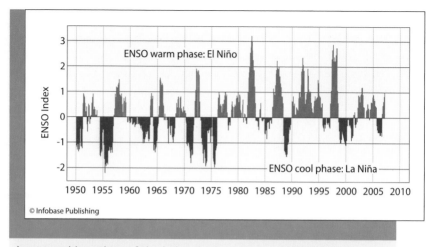

Large positive values of the index, such as during 1982–83 and 1997–98, indicate strong El Niño episodes. Negative numbers denote La Niña events. *(Source: NOAA and University of Colorado's Cooperative Institute for Research in Environmental Sciences.)*

over such broad areas. As Nash wrote in 2002, "El Niño challenges the mental capacities of humans, who only in the last few centuries have been able to understand that storms are not born just over the horizon but in the far beyond."

To measure the strength of a phase in the cycle, scientists at the National Oceanic and Atmospheric Administration (NOAA) and the University of Colorado's Cooperative Institute for Research in Environmental Sciences use a technique called the Multivariate ENSO Index. This index includes a number of variables (which makes it "multivariate"): pressure at sea level, surface winds, temperature at the surface of the ocean, and cloudiness. When the values of these measurements are averaged, the result is a number that indicates the phase and strength of the cycle. Large positive numbers occur during strong El Niño episodes, with smaller positive numbers associated with weaker events. Negative values indicate La Niña events. The figure shows the ENSO cycle from 1950 to 2008.

EFFECTS OF EL NIÑO

As shown in the figure, the Multivariate ENSO Index highlights the powerful El Niño episodes of 1982–83 and 1997–98. These years generated unusually strong effects.

The phenomenon of El Niño was initially identified by the warm current off the coast of Peru, but this is only one of its "signatures." This warm water advances northward, and during strong episodes can extend to the state of Washington and beyond, as the striped marlin—and Washington fishermen—discovered in 1997. Another effect is the change in atmospheric pressure, which reduces the Walker Circulation as well as the Pacific Ocean trade winds.

But El Niño's reach extends even farther. In 1982–83, for instance, the southwestern United States suffered from a string of severe storms that brought three or four times the normal amount of rain. As a result, flooding and landslides struck the region. In 1983, six hurricanes slammed into French Polynesia, which consists of several groups of islands in the south Pacific Ocean. These islands had not seen a hurricane in decades prior to this year. Europeans shivered in unusually frigid conditions, and droughts parched many other parts of the world, including Australia, Indonesia, China, and India.

In 1997–98, El Niño made headlines all over the world. Arid conditions in Indonesia and Malaysia led to serious forest fires, with smoke so thick that drivers could hardly see during the day. Temperatures soared in Mongolia, and parts of Africa received unaccustomed rainfall and storms. Floods inundated central Europe.

Oxford University Press released a book in 2000 called *El Niño, 1997–1998—The Climate Event of the Century,* edited by Stanley A. Changnon, in which meteorologists contributed essays on the severity and impacts of the event. Changnon wrote, "El Niño developed rapidly in the tropical Pacific during May 1997, and by October 'El Niño' had become a household phrase across

Satellite image of a hurricane—ENSO cycles have an influence on the number of storms that occur during a given year. *(NASA/Jeff Schmaltz, MODIS Land Rapid Response Team)*

America." A swarm of tornadoes erupted in the southeastern United States—Florida experienced the deadliest tornadoes in its history in late February 1998. Excessive rain caused flooding and mudslides in California, and blizzards gripped the northern portion of the Midwest.

But the 1997–98 El Niño was not without its benefits, at least in some regions. Most of the northern United States enjoyed a mild winter. Residents in this part of the country saved money because the warmer than normal temperatures lowered their heating bill.

How can people be sure that El Niño had a hand in these wide-ranging phenomena? Weather disasters such as flooding and storms, and unusual weather patterns such as exceptional mild winters, occur even in the absence of El Niño. Yet researchers looking for patterns in the world's weather have noticed that pronounced phases of the ENSO cycle are associated with an increase in weather disturbances. Some of these disturbances may be due to chance, but a preponderance of them occurring at the same time as El Niño is difficult to explain unless these events are in some way related.

Scientists use the term *teleconnection* to refer to correlations in widespread weather phenomena that are probably not due to chance. (The term *teleconnection* means a connection between distant events, and is derived from the Greek word *tēle,* which means far off. The same derivation applies to the words *television* and *telecommunication.*) Deciding if the events that happen at the same time are related or are simply the luck of the draw is not always easy. Sometimes two unrelated events occur at the same time, and sometimes an unusual abundance of events seems to happen together. A lucky gambler, for example, may enjoy a winning streak.

The branch of mathematics known as statistics helps scientists calculate expected outcomes. For example, if a coin is tossed 10 times, how many times will it land heads up? Since a "fair" coin—one that is well balanced, with both sides equally likely to turn up—has a 50/50 chance of heads, five will be the average result. But other numbers are possible. A statistical technique known as the binomial distribution shows that the chance of getting 10 heads in 10 tosses is 0.0977 percent—in other words, the odds are 1 in 1,024. If a coin-tosser consistently does better than this, people will justifiably suspect that the coin is not fair.

Meteorologists use historical weather records to determine what is "normal" activity. Statistical calculations indicate whether departures from normality are so unusual that they are probably not due to chance, as in the coin-tossing example described above.

Although two distant weather events may be related, discovering what causes their relationship is not a trivial exercise. Weather is notoriously difficult to predict more than a few days in advance, and one of the reasons is because of the complexity created by these remote connections. When separated from the rest of the system, individual components are often simpler to understand, which is why scientists generally try to isolate variables and break apart systems under study. Taken as a whole, a system of many interrelated parts is usually perplexing.

As for the teleconnections involved in El Niño, no one is sure of the details, but the basic principles are clear. The Sun's energy drives weather and atmospheric phenomena, and the Pacific Ocean, with its massive amount of water, absorbs a lot of this energy. Variations in the surface temperature of the Pacific Ocean and in the atmospheric pressure and circulation of the air above it will cause changes that ripple through the rest of the system. Adjustments in Earth's atmosphere impact weather all around the globe.

El Niño became a "household phrase" in 1997, as Changnon noted, but people must be careful not to go overboard. Strong as it might sometimes be, El Niño is not to blame for every storm, flood, or drought. Disasters happen for a multitude of reasons.

To understand any complex phenomenon such as El Niño, scientists would love to pinpoint its causes. And if the causes are known, accurate predictions will be possible. Thanks to the complexity of Earth's interacting system of ocean and atmosphere, finding the cause or causes of El Niño has been frustrating. But this has not prevented researchers and weather forecasters from issuing accurate forecasts—although El Niño packs a wallop, it takes a while for it to get wound up, giving forecasters ample warning time.

FORECASTING THE OSCILLATIONS

If an El Niño event began overnight, scientists would have a hard time seeing it coming. But the onset is quite gradual.

As discussed earlier in the chapter, atmospheric pressure drives trade winds that usually blow from east to west in the warm tropical zones around the equator in the Pacific Ocean. These winds push the warm surface water of the eastern regions toward the west. Cooler water upwelling from below replaces the loss of warm surface water, which is one of the main reasons why the waters off the coast of Peru are usually mild.

A reduction in the strength of the winds allows the warm surface water to creep back eastward. This keeps the cool water from the depths from rising. An El Niño, as illustrated in the figure on page 124, has begun.

But there is more to come. For El Niño to grow stronger, the increased warmth in the eastern region of the Pacific Ocean must become "amplified." This amplification occurs in a process known as positive feedback.

Feedback is a flow of information or energy back to its source. In engineering, feedback often occurs in electric circuits: Part of the output of the circuit, such as an outgoing current, is fed back into the input. Circuit designers sometimes employ feedback for automatic adjustment of a circuit's output, usually to lower the output if it is too high. For example, if a signal amplifier's output current is too great—which could damage the circuit components—feedback from the output turns down the amplification, resulting in a reduced and more manageable current. This is an example of negative feedback. Positive feedback arises when the output increases the amplification, but engineers usually avoid this process because it tends to produce a runaway circuit—an increase in the output causes a gain in amplification, which further increases the output, which causes an additional gain in amplification, and so on.

But nature does not shy away from positive feedback. In the beginning of an El Niño, the reduction in trade winds raises the temperature of the eastern part of the Pacific Ocean, and this rise in temperature opposes the normal circulation, which further decreases the winds. As a result of positive feedback supplied by the warm water, wind reduction creates a situation in which further decreases follow.

This process takes time to develop. For instance, early in 1997, a warming trend in the Pacific Ocean alerted marine scientists and meteorologists to the initial phase of an El Niño. Weather services issued warnings. Although no one was certain how strong the El Niño would prove to be—and it turned out to be a great deal more influential than initially expected—at least people saw it coming. Because the early stages usually appear about six months before the atmospheric adjustments and teleconnections affect weather in the United States, forecasters are able to advise Americans to prepare for storms or excessive rainfall well in advance. NOAA researcher Michael J. McPhaden, who works at the Pacific Marine Environmental Laboratory (PMEL) in Seattle, Washington, and his colleagues summarized what scientists know about the El Niño/Southern Oscillation cycle in an article published in *Science* in 2006. In the article, "ENSO as an Integrating Concept in Earth Science," the researchers noted the remarkable ability to observe and forecast the cycle's phases: "Except for the regular progression of the seasons, ENSO is the most predictable climate fluctuation on the planet."

But this predictability, once the situation begins to develop, does not mean that researchers can predict what will happen far in the future. Remember that ENSO does not have a regular cycle, as do the seasons. A variable amount of time elapses between one El Niño and the next.

One factor in this variability is the complex process of positive feedback. At some point, it must stop, but how? The termination is complicated. Waves are involved, but on a much larger scale than would be familiar to a surfer or beach-goer. An ocean is so large that it is sometimes difficult to comprehend that waves can travel from one end to another, and then bounce off a continent. The previous chapter discussed tsunamis, created during violent events such as undersea earthquakes, but other, more gentle waves exist that also travel widely (though much more slowly), and at different depths in the sea. These waves tend to offset temperature differences, returning the ocean to its original state.

Variability in the cycle is also due to the complex interactions of global wind patterns, atmospheric pressure, and the properties of the ocean—the same interactions that give rise to unpredictable weather. These variations also influence the seasons, so that one summer may not be as warm as the next. In a similar way, a cycle of ENSO may last longer or less long than the next, and one El Niño follows another after a variable period of about two to seven years.

There is also some variability in how the early stages of El Niño unfold. Although forecasters spotted the telltale signs and anticipated the 1997–98 event, the 1982–83 episode did not run the usual course—the waters off the coast of Peru began to warm in April instead of Christmas time. Roberta Conlan and Robert Service, in their 2000 paper "El Niño and La Niña," noted that after the 1982–83 El Niño, "Researchers realized that a deeper understanding of El Niño—and any hope of timely prediction—would require a much more systematic and comprehensive set of observations than were available through the programs then in operation [in the early 1980s]."

TRACKING EL NIÑO

To avoid allowing another El Niño to slip in undetected, the World Climate Research Programme, an international scientific effort, started the Tropical Ocean-Global Atmosphere (TOGA) observing system in 1985. The emphasis of this 10-year project was to collect measurements in

Tropical Atmosphere Ocean Array

The TOGA project gave scientists an opportunity to test and develop a number of monitoring devices. Stan Hayes, a researcher at the Pacific Marine Environmental Laboratory, and colleagues and technicians tinkered with an inexpensive buoy moored in the deep ocean. Deployment of these buoys continued throughout the TOGA project. Scientists evaluated the buoys' performance, fixed problems, and found satisfactory solutions. After a decade of tests involving some 400 buoys, which had been deployed on more than 80 voyages on 17 different ships, the wrinkles had been ironed out. On December 1994, the last buoy of the new array, called the Tropical Atmosphere Ocean (TAO) array, was set in place. TAO consists of a total of 70 buoys that span the tropical Pacific Ocean.

The World Climate Research Programme sponsored TOGA, but the use and maintenance of TAO shifted to various other agencies. The United States Congress voted to begin supporting TAO in 1997 as part of a long-term ENSO monitoring system. In 2000, Japanese scientists replaced some of the original equipment with their Triangle Trans Ocean Buoy Network (TRITON) devices, so TAO became a combined TAO/TRITON system. Today, the array's sponsors are the United States' NOAA, the Japan Agency for Marine-Earth Science and Technology (JAMESTEC), and the Institut de recherche pour le développement (Research Institute for Development) in France.

NOAA's National Data Buoy Center recently assumed operational control over the buoys. People interested in moni-

real time—as the events actually take place, without any data processing delay—of critical variables such as winds, water surface temperature, sea level, and the movements of waves and currents.

TOGA sent marine scientists on voyages in research vessels across the Pacific Ocean, observing and sampling the ocean's temperature and

A worker checks a TAO buoy. *(Lieutenant Mark Boland, NOAA Corps)*

toring the current conditions in the tropical Pacific Ocean can get on the Internet and visit National Data Buoy Center's TAO Web site. The Web site shows the buoy locations and provides data plots of wind speed, sea surface temperature, depth averaged temperature, air temperature, relative humidity, and more.

motion. Researchers also collected data from moored buoys, which contained sensors and instruments. The buoys transmitted the data to observatories, often via satellite links. This network kept scientists updated with fresh, real-time information about temperatures, sea level, currents, wind direction, and speed, from all over the ocean. From this information, oceanographers and meteorologists constructed models of oceanic and atmospheric phenomena related to ENSO. These models, which were incorporated into computer programs, used the values of important variables to predict El Niño's course.

As TOGA wound down, scientists took the lessons learned from this project and built the monitoring system known as the Tropical Atmosphere and Ocean (TAO) array. As described in the sidebar, this system consists of a network of buoys that keep a close watch on the conditions in tropical waters.

Information from this array of sensors helped oceanographers and meteorologists detect the early pressure and temperature variations of the 1997–98 El Niño, as well as other, less intense episodes that have occurred more recently. Forecasters believe they will continue to be able to spot future El Niño events in the earliest stages.

But researchers would like to use this data, and other observations, to extend their investigation of ENSO even further. Learning more about this major weather-maker might permit long-range forecasts of the cycle, rather than just detecting when a certain phase, such as El Niño, has begun. Such discoveries would also enhance scientific knowledge of oceans, atmosphere, and weather. One of the most critical points that needs to be studied is the nature of the factor or factors that drive the ENSO cycle.

CAUSES OF EL NIÑO

Although researchers know how El Niño evolves once it gets started, they do not know what causes it to get started. The previous sections of this chapter described the early stages of El Niño, which begins when the atmospheric pressure in the tropical Pacific Ocean changes and the trade winds decrease, followed by a warming of the sea surface temperature off the coast of Peru and the eastern Pacific Ocean, which further decreases the trade winds and enhances the rise in temperature. But scientists do not yet know what kicks off this progression of events.

The main obstacles in this line of research are those ocean-air interactions, such as the feedback loop described earlier. Winds affect the ocean temperature, which affects the winds, and this, in turn, affects the ocean temperature. Parts of this complicated system are well understood; for example, meteorologists have a good handle on localized atmospheric pressure and winds, as well as heat flow and evaporation. The ocean tends to remain steady over short periods of time, since sea temperatures change slowly. But over the long term, scientists do not understand the complicated interactions between the ocean and the atmosphere. These interactions are global in scale and involve many different subsystems. Interdisciplinary effort is required to tackle these problems—meteorologists and oceanographers must cooperate, since the pieces of this puzzle fall into both specialties.

What initiates the reduction of trade winds at the beginning of an El Niño? Theories are beginning to emerge. One theory entertains the notion that the tropical Pacific Ocean, and the air above it, behaves like an oscillator, naturally swinging back and forth between El Niño and La Niña phases. No special event is needed to initiate the process—the cycle occurs because the ocean-atmosphere system oscillates. A pendulum, for example, swings from side to side, once it is set into motion, as long as there is a driving force, such as the falling weights in a grandfather clock. In the oscillation theory, ENSO is like a pendulum, set into motion long ago, and driven by the Sun's energy. But unlike a pendulum, ENSO does not go through its cycle in a fixed period of time. This variability does not ruin the theory, but it indicates that other interactions or factors come into play. The identity of these interactions or factors remains unknown.

Some researchers have argued that the main oscillation is due to one or just a few factors. In this view, better predictions of El Niño will result from models that can quickly spot trends in these factors, so that the earliest phases of the oscillation's cycle can be identified. Dake Chen, a researcher at Columbia University in New York, and his colleagues recently used sea surface temperature data from the period 1857–2003 to train a computer model to detect subtle cues presaging El Niño events. Previous models could identify trends a few months before El Niño gets started, but the model of Chen and his colleagues did better than this. "The model successfully predicts all prominent El Niño events within this period [1857–2003] at lead times of up to two years," wrote the researchers in their report, "Predictability of El Niño over the Past 148 Years," published in a 2004 issue of *Nature*.

An alternative theory posits that the ocean-atmosphere system in the Pacific Ocean is generally stable rather than oscillatory. El Niños occur when something triggers a cycle, in which case the system goes through its phases, and then waits for another trigger. In this theory, the irregularity of the trigger accounts for ENSO's variability.

The success of the model of Chen and his colleagues supports the theory that ENSO is an oscillator. If an outside force is necessary to trigger the cycle, then predictions based on sea surface temperature would probably be unable to predict an El Niño very far in advance—one would have to know the trigger in order to do that.

Yet the issue is far from decided. The data used by Chen and his colleague may reflect the influence of the trigger. At the present time, no one is certain if ENSO is a self-sustained oscillation, needing only the Sun's energy to go through its cycle, or if the system is relatively stable and requires some sort of disturbance to trigger a cycle. The identity of this disturbance, if it exists, is unknown.

Both theories may even turn out to be true. Although it is not possible that both theories are correct for the same period of time, either theory may hold true for certain intervals. For instance, the system may oscillate under certain conditions, and be stable in others. When the oscillatory conditions prevail, the system needs no trigger. At other times, cycles do not occur unless triggered. What these conditions are, and how they might cause oscillatory or stable behavior, is not clear.

The only clear thing concerning the causes of El Niño is that more research needs to be done. Large scale phenomena such as ENSO make observations difficult and require extensive and complex computer models, so researchers have their work cut out for them. Yet considering the impact these events have had on Earth's weather systems and climate in the past and present—and their potential impact in the future—this research is important.

EL NIÑO AND EARTH'S WEATHER AND CLIMATE

Earth's surface temperature has warmed about 1.33°F (0.74°C) in the last century according the Intergovernmental Panel on Climate Change. Although the magnitude of this global climate change may not seem like much, it can have major effects while upsetting the delicate balance of Earth's ocean and atmospheric systems. Consider ENSO, for example—these tropical Pacific Ocean cycles are often powerful enough to influence weather across the world, yet the shifts or transitions between phases are generally subtle.

No one can predict with any certainty the future course of the complex set of interacting systems that comprise Earth's weather and climate. Yet the impact on society could be tremendous, as changing conditions may significantly alter sea levels, agriculture, and water distribution. Scientists who study climate and weather need as much information as they can get to make their computer models more accurate. Phenomena such as El Niño, which can have such widespread effects, must be taken into consideration.

Events in the tropical Pacific Ocean clearly interact with and have a major impact on worldwide weather systems. What happens to ENSO if Earth continues to get warmer? This is an important but unanswered question, since researchers do not yet have a sufficient amount of data or an adequate understanding of climate to solve the issue. As Richard A. Kerr, a science writer, observed in *Science* on July 29, 2005, "Computer climate models disagree about how future global warming will affect it [the tropical Pacific Ocean]: whether the region will get stuck in the warmth of a permanent El Niño, slip into the relative cool of an endless La Niña, or keep swinging from one to the other as it does today." Whatever happens will have a powerful effect on the world's weather.

In order to try to answer this question, some climatologists look to the past. Parts of the Pliocene Epoch, which began about 5.3 million years ago and lasted until 1.8 million years ago, were quite warm, several degrees above today's temperature. If scientists could discover how the tropical Pacific Ocean behaved under these conditions, its future behavior may become more predictable.

But events millions of years in the past are not easy to discern. Some researchers have analyzed the composition of marine fossils from sediments of the Pliocene Epoch, hoping to get some clues regarding the temperature of the water in which these creatures lived. For example, the magnesium/calcium ratio in the shells of certain sea-dwelling organisms reflects, at least to some extent, the temperature of the water. These shells are mostly made of calcium carbonate, but a small number of atoms of magnesium are included, and this number tends to increase at higher temperature. But this kind of data is subject to many errors.

For instance, researchers can only locate and measure the fossils of a small sample of the organisms that lived during this epoch, and there is no way of being sure that this sample is typical of the whole population. If it is not, then the results are not valid in general. The clues about the temperature and state of the ocean during the Pliocene Epoch have yielded mixed results rather than a firm answer.

The complexities of the climate have frustrated efforts to probe its past as well as predict its future. But an interesting aspect of the ENSO cycle is that prominent episodes have predictable effects. This predictability greatly improves the accuracy of advance weather forecasts, such as predicting the severity of the coming winter or the amount of expected precipitation, since these forecasts otherwise often turn out to be wrong. Writing in the August 15, 2008, issue of *Science,* Kerr notes that "without an El Niño or its counterpart, La Niña, next winter's weather is pretty much anybody's guess." El Niño sometimes gives meteorologists a helping hand.

As Earth's climate continues to change, many researchers worry that the planet and some of its oceanic and atmospheric systems may be approaching a state in which the accumulative effect could be drastic. Like the straw that broke the camel's back, even a slight addition to an already fully loaded system can cause a significant change. Timothy M. Lenton, a researcher at the University of East Anglia in the United Kingdom, and his colleagues published a paper in 2008 in the *Proceedings of the National Academy of Sciences* called, "Tipping Elements in the Earth's Climate System." The authors note, "The term 'tipping point' commonly refers to a critical threshold at which a tiny perturbation can qualitatively alter the state or development of a system." By "qualitatively alter," the researchers mean an entirely different state or mode of operation, rather than just a minor adjustment of the present state.

In the paper, the researchers report their efforts to identify the components or systems of Earth's climate that are most likely to experience drastic changes in the near future. Lenton and his colleagues reviewed data and ideas published in science journals as well as the discussions at a conference of 36 experts in Berlin, Germany, in 2005, and specified nine systems that they believe are most susceptible to being "tipped" into a different state within the next 100 years or so. Some of these systems involve ice packs, such as the Greenland ice sheet and the West Antarctic ice sheet, which could melt and increase the sea level; other systems involve trees, such as in the Amazon rain forest, the deforestation of which would significantly reduce biological diversity and precipitation. ENSO also made the list. Lenton and his colleagues consider this system at a "tipping point," which may swing the oscillation into one extreme state or another, or cause the oscillations to shift more violently than before.

CONCLUSION

Pinpointing the cause or causes of El Niño, and predicting the future behavior of the ENSO cycle, will require much more research. Observations and measurements in this frontier of marine science need a global perspective. This is why researchers welcome the involvement of the National Aeronautics and Space Administration (NASA), the United States government agency devoted to space science and exploration. For a global perspective, nothing beats the view of an orbiting satellite.

Chapter 1 described how satellites measure the contour of the ocean surface with great precision, which reveals prominent features of the seafloor below. Fluctuations in sea height are also important, since these variations often arise when winds drive water from one place to another or when changing temperatures cause the water to expand or contract (this is an example of thermal expansion—objects tend to expand when heated, and contract when cooled). Temperature and winds are both critical in El Niño and La Niña.

Jason-1, a satellite launched on December 7, 2001, carries an altimeter that can measure sea level to an accuracy of 1.3 inches (3.3 cm). The satellite, which was named after the mythological Greek mariner, orbits at an altitude of 825 miles (1,330 km). Researchers analyze this data to keep track of conditions in the tropical Pacific Ocean and elsewhere. In 2007, for example, *Jason-1* reported a less than normal sea level in the eastern part of the tropical Pacific Ocean. These observations coincided with a La Niña episode, in which strengthened winds blow the warm surface water toward the west, and cool water from below takes its place.

On June 20, 2008, *Jason-2* made a successful launch, achieving orbit at the same altitude as *Jason-1.* This satellite, designed and built by several space agencies including NASA, contains advanced altimetry instruments as well as other sophisticated sensors. Scientists hope the

Depiction of a satellite monitoring the ocean *(NASA)*

sensitive instruments on this craft can improve the accuracy of sea level measurements even further, perhaps down to 1 inch (2.5 cm).

Satellites can also monitor surface temperatures on land and sea. NASA's *Terra* satellite, launched on December 18, 1999, orbits Earth at an altitude of 437 miles (705 km). Among this satellite's instruments is the Moderate Resolution Imaging Spectroradiometer (MODIS), which detects and analyzes radiation from land and sea surfaces. The amount and frequency distribution of this radiation depends on the properties of the surface, including its temperature.

Another NASA satellite, *Aqua,* was launched on May 4, 2002. *Aqua* orbits at the same altitude as *Terra,* and also carries the MODIS package. The orbits of these two satellites are timed and angled so that the MODIS instruments see the whole surface of the planet every few days. MODIS not only measures surface temperatures, it also detects the amount of *phytoplankton* in the water, based on the effects these photosynthetic marine organisms have on the spectrum of radiation. By combining temperature and phytoplankton measurements, as well as other data, MODIS will indicate the extent to which El Niño and La Niña affect many of the ocean's properties, including its life-forms.

Satellite data, coupled with TAO data and other observations, which researchers obtain during lengthy voyages in research vessels, provides scientists with a more complete picture of the conditions and events in the tropical Pacific Ocean. As researchers collate and analyze this mountain of data, they will get a better idea of the nature of ENSO's cycles, and perhaps what causes them. In the 2006 *Science* paper of McPhaden and his colleagues, cited earlier, the authors wrote that "many intertwined issues regarding ENSO dynamics, impacts, forecasting, and applications remain unresolved. Research to address these issues will not only lead to progress across a broad range of scientific disciplines but also provide an opportunity to educate the public and policy makers about the importance of climate variability and change in the modern world."

Ever since the severe 1997–98 El Niño episode, which made the term *El Niño* a household phrase, interest in this phenomenon seems to have diminished—fewer households are talking about it, and fewer articles are appearing in the journals. This is understandable, since El Niño has faded from the news headlines. But another severe episode will occur in the future, and then another, as they have throughout history. People need a better understanding of these phenomena to alleviate the adverse impacts that will undoubtedly come from these events. Advances in this frontier of marine science will also deliver critical knowledge about Earth's oceans and climate, which will be of great help as people face, and try to overcome, challenges such as global climate change.

CHRONOLOGY

1519–21   The around-the-world voyage of Ferdinand Magellan (1480–1521) and his men may have been strongly impacted by an El Niño event, according to the research of Scott M. Fitzpatrick and Richard Callaghan.

1800s   South American sailors and fisherman use the term *El Niño* to specify the unusually warm water that sometimes appears during Christmas off Peru's coast.

1891   A particularly strong El Niño off the coast of Peru stimulates scientific interest in the phenomenon.

1924   British researcher Sir Gilbert Walker (1868–1958) publishes his discovery of the Southern Oscillation.

1925   Another strong El Niño piques researcher interest.

1957–58   The International Council of Scientific Unions designates this period as the International Geophysical Year, during which scientists from all over the world made important observations, including those related to the tropical Pacific Ocean and an El Niño that was occurring at this time.

1960s   UCLA scientist Jacob Bjerknes (1897–1975) links El Niño with the Southern Oscillation

1982–83   A particularly strong El Niño attracts widespread scientific interest in the phenomenon.

1985–94   The World Climate Research Programme, an international research effort, conducts the Tropical Ocean-Global Atmosphere (TOGA) observing system.

1994   The final buoy of the 70-buoy Tropical Atmosphere Ocean (TAO) array gets moored.

1997–98	A powerful El Niño alters weather across the globe and garners worldwide attention.
2000–08	Weak to moderate El Niño and La Niña episodes occur.
2004	Columbia University researcher Dake Chen and his colleagues train a computer model to use sea surface temperature data from the period 1857–2003 to spot early phases of El Niño events, up to two years before conditions intensify.

FURTHER RESOURCES
Print and Internet

Changnon, Stanley A., ed. *El Niño, 1997–1998—The Climate Event of the Century*. Oxford: Oxford University Press, 2000. This set of essays focuses on one of the strongest El Niño episodes in recent memory. Topics include the key events of the phenomenon, causes and predictions, scientific issues, and the impact of El Niño on future weather patterns.

Chen, Dake, Mark A. Cane, Alexey Kaplan, Stephen E. Zebiak, and Daji Huang. "Predictability of El Niño over the Past 148 Years." *Nature* 428 (April 15, 2004): 733–736. The researchers use sea surface temperature data to train a computer model to detect subtle cues presaging El Niño events.

Conlan, Roberta, and Robert Service. "El Niño and La Niña: Tracing the Dance of Ocean and Atmosphere." Available online. URL: http://www7. nationalacademies.org/opus/elnino.html. Accessed June 9, 2009. This article, published by the National Academy of Sciences in 2000, describes the evolution of knowledge concerning El Niño and La Niña. The work of Gilbert Walker, Jacob Bjerknes, and others is discussed.

Fagan, Brian M. *Floods, Famines, and Emperors: El Niño and the Fate of Civilizations*. New York: Basic Books, 1999. Science writer and anthropologist Brian Fagan takes a societal perspective in this book, examining how people commonly deal with disruptions from El Niño and similar weather events.

Franklin Institute. "El Niño: Hot Air over Hot Water." Available online. URL: http://www.fi.edu/weather/nino/nino.html. Accessed June 9, 2009. The Franklin Institute, a science museum located in Philadelphia, Pennsylvania, adopts a hands-on approach to explaining science. This Web resource describes some simple experiments that can help people gain a better idea of how El Niño works.

Kerr, Richard A. "El Niño or La Niña? The Past Hints at the Future." *Science* 309 (July 29, 2005): 687. Kerr writes about the uncertainty of what these weather phenomena may entail in the future.

———. "Seasonal-Climate Forecasts Improving Ever So Slowly." *Science* 321 (August 15, 2008): 900–901. Kerr notes the gradual improvement in climate forecasts.

Lenton, Timothy M., Hermann Held, Elmar Kriegler, Jim W. Hall, Wolfgang Lucht, Stefan Rahmstorf, et al. "Tipping Elements in the Earth's Climate System." *Proceedings of the National Academy of Sciences* 105 (2008): 1,786–1,793. The researchers enumerate thresholds on which the climate may swing drastically in one direction or another.

McPhaden, Michael J., Stephen E. Zebiak, and Michael H. Glantz. "ENSO as an Integrating Concept in Earth Science." *Science* 314 (December 15, 2006): 1,740–1,745. This review summarizes what scientists know about the El Niño/Southern Oscillation cycle.

Nash, J. Madeleine. *El Niño: Unlocking the Secrets of the Master Weather-Maker.* New York: Warner Books, 2002. Science writer J. Madeleine Nash describes these events from a scientific perspective as well as a humanistic one. She tells the stories of scientists who study these powerful phenomena and the people who experience them and are strongly, and often adversely, affected.

Pacific Marine Environmental Laboratory. "What Is an El Niño?" Available online. URL: http://www.pmel.noaa.gov/tao/elnino/el-nino-story.html. Accessed June 9, 2009. With illustrations and animations, this Web page explains how researchers recognize when an El Niño is happening, and the effects these events can have.

Philander, S. George. *Our Affair with El Niño: How We Transformed an Enchanting Peruvian Current into a Global Climate Hazard.* Princeton, N.J.: Princeton University Press, 2004. This book explains what scientists know about this weather phenomenon, as well as recounting the history of people's perceptions of El Niño and the changes wrought by the strong event of 1997–98.

Public Broadcasting Service. "Tracking El Niño." Available online. URL: http://www.pbs.org/wgbh/nova/elnino/. Accessed June 9, 2009. This Web resource is a companion to an episode of the science series NOVA, aired in 1998, and is full of information about how El Niño gets started and how its reach extends to weather systems all over the world.

Suplee, Curt. "El Niño/La Niña." *National Geographic.* March 1999. Available online. URL: http://www.nationalgeographic.com/elnino/mainpage.html. Accessed June 9, 2009. This article covers what researchers have learned about these weather phenomena.

Web Sites

National Aeronautics and Space Administration: Earth Observing System. Available online. URL: http://eospso.gsfc.nasa.gov/. Accessed June 9, 2009. This Web site contains information on NASA's missions, such as the *Terra, Aqua,* and *Jason* satellites, dedicated to observing and studying Earth.

National Data Buoy Center: Tropical Atmosphere Ocean Array. Available online. URL: http://tao.noaa.gov/. Accessed June 9, 2009. Plots of current and past TAO data are available. The Web site also includes information on the status of the buoys, a "virtual" tour, and a history of the array's development.

National Oceanic and Atmospheric Administration: El Niño Page. Available online. URL: http://www.elnino.noaa.gov/. Accessed June 9, 2009. Updates of the present state of the tropical Pacific Ocean are presented here. There are also links to many pages discussing El Niño forecasts, observations, and research.

6

HARMFUL ALGAL BLOOMS—"RED TIDES"

Many people who spent their 2005 vacations on the beaches of the Gulf of Mexico did not have the best of times. Randolph Fillmore, writing in *USF Magazine,* a publication of the University of South Florida, described in the winter 2005 issue the unpleasantness that year: "Piles of beached dead fish. Beach-going tourists coughing in the toxic air. The smelly cleanups." The cause of the disaster was a phenomenon known as a red tide, which began "killing off fish and birds, striking a serious blow at the fishing and tourism industries and doing its best to ruin paradise as we know it."

Red tides tend to discolor the water, which is the source of their name. But red tides have nothing to do with tides, and they are not always red either—they can be green, brown, or have no effect on the water's color at all. Scientists prefer to call them *harmful algal blooms,* because the responsibility rests on small organisms known as algae, whose population occasionally increases dramatically, like a sudden bloom. Algal blooms occur often and many have little or no environmental consequences, but some species of algae release poisons and create havoc. Florida and other coastal states have been suffering from a lot of harmful algal blooms recently, some of which have been prolonged and deadly. According to the article in *USF Magazine,* "Marine biologists point out that while red tide occurs naturally every year in the Gulf of Mexico, the 2005 red tide will likely go down in the books as one of the worst and longest duration as the Gulf coast region suffered the effects of red tide for months on end."

This chapter describes some of the ill effects of harmful algal blooms. Harmful algae produce poisonous substances, known as *toxins,* which get into the food chain when consumed by marine animals. Animals

Algal bloom at the Bountiful Islands of Australia *(Bill Bachman/Photo Researchers, Inc.)*

such as birds and humans get sick when they eat contaminated seafood. Diseases such as *paralytic shellfish poisoning* cause serious illness and even death in humans. Another adverse consequence of certain algal blooms involves a disruption in the normal ecology of coastal waters. Massive increases in algae populations sometimes invoke a corresponding increase in organisms that feed off algae, such as certain kinds of bacteria whose voracious metabolic activity depletes the oxygen supply. The resulting condition is known as *hypoxia*—little oxygen is available for fish and other marine creatures, creating a *dead zone.*

The increasing severity of harmful algal blooms prompted the United States government to pass the Harmful Algal Bloom and Hypoxia Research and Control Act in 1998. Officials noted that "harmful algal blooms may have been responsible for an estimated $1,000,000,000 in economic losses during the past decade" and called for more monitoring, assessments, and research. In response to these challenges, marine scientists have been investigating how harmful algal blooms arise and the best methods to monitor, alleviate, and perhaps even prevent their occurrence.

INTRODUCTION

The oceans contain an abundance of life. At the bottom of the scale are tiny organisms that float and drift with the currents. Many of these organisms are microscopic plants called algae.

Algae are a diverse group of plant life. They are photosynthetic, which means they use the energy of sunlight to produce carbohydrates. About 35,000 species of algae are known to science. Most are aquatic, found in freshwater or salt water. The need for sunlight limits their habitat to the surface or shallow depths, since light does not penetrate deeper waters. They are usually simple organisms, although some can be quite large—seaweeds such as kelp are classified as algae, and can exceed 164 feet (50 m) in length.

Microscopic species of algae are the kind that occasionally experiences a population explosion known as a bloom. Even in periods of normal population, these tiny, single-celled plants are numerous—every drop of water teems with them. Algae compose a large component of plankton, the drifting organisms which are a critical source of food in the oceans, and are the foundation of the marine food chain. Given enough sunlight and a healthy marine environment, algae multiply rapidly, providing food for other microorganisms as well as fish and whales.

Most algae reproduce by a process known as asexual fission. Under conditions of plentiful sunlight and nutrients, the cell grows until it reaches a point at which it fissions, or divides, into two cells, as shown on the left side of the figure. These cells grow and eventually divide, producing other cells, and the process continues as long as the sunlight and nutrients hold out. But nature does not leave this growth process unchecked, since predators abound in the seas, and single-celled algae sit at the bottom of the food chain.

Some species of algae, such as members of the genus *Alexandrium*, can enter another phase of their life cycle that greatly contributes to their survival. In times of stress, when sunlight and nutrients are scarce,

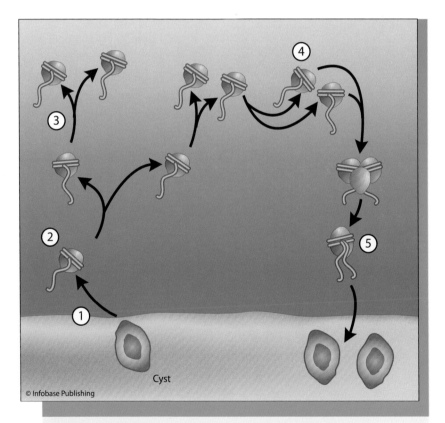

Asexual reproduction consists of a growth stage (2) and division (3). During harsh conditions, some species of algae fuse (4) to form a zygote (5), which shuts down its metabolism and builds a protective wall, forming a cyst (1).

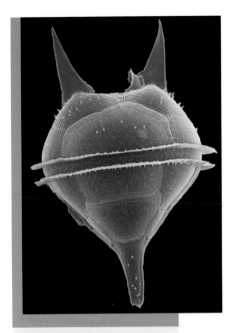

Greatly magnified image of the dinoflagellate *Ceratium furca* (Steve Gschmeissner/ Photo Researchers, Inc.)

the cells switch to sexual reproduction. As illustrated in the right side of the figure, two cells fuse—the opposite of fission—and produce a cell known as a zygote. This cell builds up a thick wall to protect itself and goes into a dormant stage. In this stage, the organism is called a *cyst*. Cysts settle on the bottom and can survive for a number of years under harsh conditions; the cysts can even endure ingestion and a trip through a predator's digestive system.

Dormancy is a mechanism similar to hibernation, by which organisms manage to live through hard times by shutting down key physiological processes. When conditions favorable to life return—for instance, the presence of sunlight and warm water—the cell breaks out of its protective wall and begins to grow, starting another cycle of asexual reproduction. The dormant stage is also an excellent means of finding a new place to live. Instead of settling on the bottom, some cysts may be carried by strong currents, ingested by a fish, or get sucked into the ballast tank of a cargo vessel. (In order to maintain stability, a captain will fill the ship's ballast tanks with water when the vessel is not carrying a lot of cargo.) Hitching a ride on one or more of these modes of travel, cysts can disperse widely, finding new and richer habitats in which to grow. When the conditions are ripe, a bloom begins.

Some of the most common "blooming" algae in the ocean belong to a group of organisms known as *dinoflagellates*. These organisms move about by flailing thin, whiplike appendages known as flagella. (The term *dinoflagellate* comes from a combination of a Greek word, *dinos,* meaning "rotating," and a Latin word, *flagellum,* which refers to a whip.) *Alexandrium* species, for example, are dinoflagellates.

An algal bloom in and of itself is not reason for great concern. And for certain predators, it is a time to feast. But under certain circumstances, an algal bloom can be a major hazard.

ALGAL BLOOMS THAT CAUSE TROUBLE

"Red tides" have been known throughout history, although early scientists and naturalists were unaware of the organisms that produced them. Spanish explorer Álvar Núñez Cabeza de Vaca (ca. 1490–1557) visited the Gulf of Mexico in the 16th century and reported seeing a red tide, and noted that such phenomena occur every year, killing many fish. Some people have also speculated that the first plague of Egypt, as mentioned in the biblical chapter of Exodus, refers to a red tide: "And all the waters that were in the river were turned to blood. The fish that were in the river died, the river stank, and the Egyptians could not drink the water of the river."

Although the phenomenon has nothing to do with the tides of the ocean, the presence of so many algae sometimes colors the water, and the bloom seems to roll in like a tide. But the color of the water depends on the species and sometimes also on the concentration. *Karenia brevis* is a dinoflagellate whose blooms afflict the coast of Florida, turning the water a reddish-brown color, as do some species of *Alexandrium*. But other species may turn water various shades of brown or green, or they may not color the water at all.

Algal blooms are not limited to the oceans. Blooms also occur in freshwater algae or certain types of photosynthetic bacteria known as cyanobacteria. Cyanobacteria are aquatic and engage in photosynthesis, and under certain conditions form a blue-green scum on ponds, so these organisms are sometimes called blue-green algae. The scope of this chapter is limited to marine algae, although much of the discussion applies to other organisms.

Population explosions of algae cause problems when the blooms deprive other creatures of nutrients or produce high concentrations of harmful substances. These harmful blooms occur in some form or another in all parts of the world. Writing in a 2004 issue of *Oceanus,* the magazine of the Woods Hole Oceanographic Institution (WHOI), WHOI researcher Donald M. Anderson noted, "Virtually every coastal country is threatened by multiple harmful or toxic algal species. Bloom

New England "Red Tide" of 1972

Canada and parts of Maine had experienced periodic outbreaks of harmful algal blooms that caused paralytic shellfish poisoning in people, but Massachusetts had been relatively free of the disease until 1972. In autumn of that year, a strong storm altered wind and current patterns in the area, driving *Alexandrium* dinoflagellates that normally thrive in Canadian waters farther south. During the month of September, a harmful algal bloom of massive proportions struck, stretching from southern Maine into New Hampshire and parts of Massachusetts. The bloom was visible as a reddish discoloration of the water—a "red tide" worthy of the name.

Several dozen cases of paralytic shellfish poisoning occurred, and government officials enacted a temporary ban on harvesting seafood such as clams, mussels, and oysters that sometimes store the toxin that causes the disease. The drastic measure idled fishermen and drained millions of dollars from the local economy, as well as frightening restaurant patrons into avoiding all kinds of seafood, even safe dishes.

The 1972 harmful algal bloom was especially important because it served to introduce the algae into the area—and they stayed, in the form of cysts. As a result, periodic outbreaks have occurred almost every year in Maine, and less frequently in New Hampshire and Massachusetts. Most outbreaks are relatively minor, but in 2005, a harmful algal bloom broke out that was big enough to rival that of 1972. Possible contributors to the severity of the 2005 bloom were a couple of storms—nor'easters, as they are called in the region, because the wind blows in from the northeast—and a warmer spring and summer. The New England fishing industry was so hard hit in 2005 that Mitt Romney, then governor of Massachusetts, requested and received federal disaster aid.

events can cover areas as small as a coastal pond or as large as one million football fields."

The number of harmful algal blooms may be increasing recently, although it is difficult to be sure since historical records are spotty at best. But some areas have definitely started to experience "red tides" where few such episodes have occurred in recent memory. For example, blooms associated with paralytic shellfish poisoning were practically unknown in Massachusetts prior to a harmful algal bloom developed in 1972. The sidebar provides more details on this event.

Paralytic shellfish poisoning results from consumption of algae toxins. High concentrations of these toxins are one of the worst effects of harmful algal blooms.

TOXINS OF HARMFUL ALGAE

A toxin is a poisonous substance. *Alexandrium* and other species produce water-soluble toxins such as saxitoxin and similar substances that accumulate in the tissues of shellfish as they consume dinoflagellates. Shellfish may be exposed to a negligible amount of toxin during normal periods, but they may acquire significant amounts during a bloom. Many of these toxins do little damage to the shellfish, but cause a lot of problems in fish, marine mammals, and humans that eat the contaminated shellfish. Toxicity is not diminished by cooking or freezing.

Saxitoxin, for example, attacks certain proteins important for the conduction of nerve and muscle impulses, causing paralytic shellfish poisoning. In humans, the symptoms begin with tingling of the extremities, loss of movement in arms and legs, and labored breathing. Depending on the amount of toxin ingested, the victim may experience paralysis of the muscles of the chest and abdomen. In severe cases, respiratory arrest and death may ensue, although these occurrences are rare.

Other toxins may follow a different route through the food chain. The figure shows some of the possible processes. Contamination may spread from microbes to fish larva, shellfish, fish, seals, sea birds, and humans. Such poisonings are probably responsible for ancient taboos in some early Native American cultures against eating shellfish during certain times. Harmful algal blooms are especially prevalent in the months of April through September in the Northern Hemisphere.

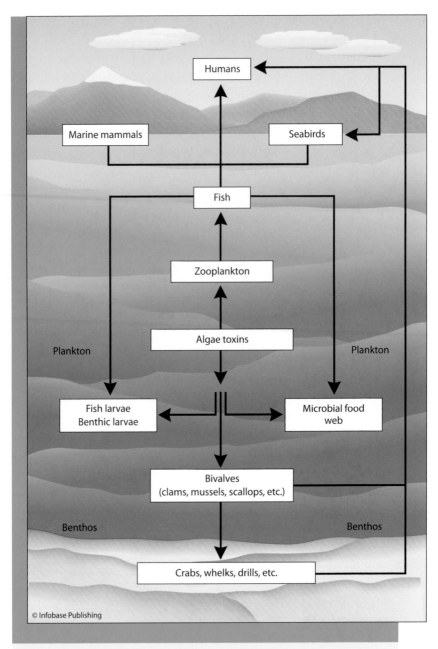

Algal toxins can enter the food chain in a variety of ways, eventually reaching humans.

Few cases of paralytic shellfish poisoning or other maladies associated with toxic algae happen in the United States, due to the diligence of officials and fishermen who are on the lookout for harmful algal blooms. When a bloom is in progress, fishermen avoid shellfish and other possibly contaminated organisms. Shellfish usually eliminate any absorbed toxins after a short period of time. Keeping a close watch on harmful algal blooms is easy when the algae discolor the water, but in other cases, the signs are not so obvious. Techniques that provide adequate warning are discussed in the section "Monitoring Algal Blooms" later in this chapter.

Why does an otherwise unassuming single-celled organism produce such a toxic substance? Some animals, such as rattlesnakes, use venom for defense and capturing prey; other animals, such as certain species of small frogs, produce toxins to discourage predators (who are usually warned by bright, distinctive colors of the toxic prey). In the case of toxic microscopic algae, neither situation seems to apply.

No one knows why certain algae produce toxins, but one theory holds that these substances are used in special biochemical reactions necessary for the organism's *metabolism*. If this theory is true, then the toxicity of these substances to other organisms is a "side effect"—an unintentional consequence. More research on algal biochemistry will be needed to address this issue.

DEAD ZONES—NOT ENOUGH NUTRIENTS FOR LIFE

Only about 100 or so species of algae produce toxic substances. But even the blooms of poison-free algae can create problems. In some cases, algal blooms harm the environment by crowding out other life. Some of these harmful algal blooms get a helping hand from humans, albeit accidentally.

Agriculture in the United States and many other countries has become more productive for a variety of reasons, including the extensive use of fertilizers. The addition of nutrients into the soil boosts crop yields, but some of the chemicals are not absorbed, and wash away in the rain. Other activities, such as the use of certain fuels, also contribute to chemical runoff. In a 2004 issue of *Oceanus,* WHOI scientist Andrew R. Solow wrote, "The most widespread, chronic environmental problem in the coastal ocean is caused by an excess of chemical nutrients. Over the past century, a wide range of human activities—the intensification of agriculture, waste disposal, coastal development, and fossil fuel use—has substantially increased the discharge of nitrogen, phosphorus, and other nutrients into the environment. These nutrients are moved around by streams, rivers, groundwater, sewage outfalls, and the atmosphere and eventually end up in the ocean."

For marine microorganisms, the extra nutrients flowing into coastal waters provide a temporary feast. This situation is called *eutrophication,* a term meaning well-nourished. Algae bloom readily in the rich water. But the good times do not last. Following the feast is a famine, as the nutrients are unable to keep up with the rapid rise in population. Billions of algae die and sink to the bottom. This massive amount of dead algae becomes a feast for bacteria, and in the process, the bacteria use up most of the oxygen dissolved in the water. This problem does not become too severe if the water is well mixed, allowing oxygen from other regions to replenish the depleted area. But many coastal waters are not well mixed, especially in bays, gulfs, and protected areas. If rapid mixing does not occur, oxygen may drop to 40 percent or less of its normal level, which is too low for most fish and other marine life to survive. Animals either swim away or die. The result is a dead zone lasting until enough oxygen finally seeps into the area, eliminating the hypoxia.

"Hypoxia occurs throughout the world," wrote Solow in his 2004 article. "Two of the best-known hypoxic areas are in the Black Sea and the Baltic Sea. In the U.S., dead zones occur regularly in Long Island Sound, the Chesapeake Bay, and the northern Gulf of Mexico. In the Baltic Sea, hypoxia has contributed to the collapse of the Norwegian lobster fishery. There is evidence that the hypoxia off the coast of Louisiana has harmed the valuable shrimp fishery and possibly contributed to the replacement of bottom-dwelling species such as snapper with less valuable mid-water species such as menhaden." A lot of the problems off the Louisiana coast are due to the Mississippi River, which drains a great deal of agricultural lands in the country's midsection.

TRIGGERING RUNAWAY GROWTH

Some scientists believe that the huge outflow of the Mississippi River, the largest river system in North America, may be contributing to algal blooms associated with *Karenia brevis* and related species all the way to Florida. NOAA scientist Richard P. Stumpf and his colleagues recently promoted this hypothesis in a 2008 paper, "Hydrodynamic Accumulation of *Karenia* off the West Coast of Florida," published in *Continental Shelf Research.*

The researchers used satellite imagery that measured the temperature and color of surface water. This data indicated a high concentration of algae at the continental shelf along the west coast of Florida. Samples from these areas show *Karenia* populations vastly exceed what would be expected even if the algae were growing at a maximum rate. Stumpf and his colleagues proposed a model to account for high *Karenia* populations, even when the local supply of nutrients should not be able to support them. In the model, summer winds carry water and nutrients from the Mississippi River outflow onto the west Florida continental shelf. According to Stumpf and his coworkers, "This water mass supplies utilizable inorganic and organic forms of nitrogen that promote the growth of *Karenia* to pre-bloom concentrations in sub-surface waters in the mid-shelf region. In the fall, a change to upwelling favorable winds produces onshore transport. This transport, coupled with the swimming behavior of *Karenia*, leads to physical accumulation at frontal regions near the coast, resulting in fall blooms."

Mississippi River nutrients are widely believed to contribute to algal blooms in Texas, as prevailing currents tend to push the water westward. Although a temporary change in conditions could steer the water toward Florida, as Stumpf and his team suggest, there is a question of how long the nutrients travel before they are consumed or settle down to the bottom. In an article written by Craig Pittman and published in the *St. Petersburg Times* on November 8, 2007, Florida Fish and Wildlife Research Laboratory scientist Cindy Heil said that her research suggests that these waters are "generally stripped of nutrients" before reaching Florida. NOAA is attempting to test these opposing theories by using underwater vehicles to sample *Karenia* populations throughout the Gulf of Mexico. If there is a "trail" of nutrients, perhaps scientists can locate it.

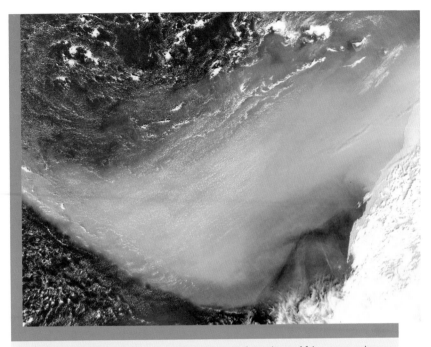

Satellite image of a dust storm off the coast of northern Africa, sweeping out over the Atlantic Ocean *(Jacques Descloitres, MODIS Rapid Response Team, NASA/GSFC)*

Another hypothetical source of harmful algal blooms in the Gulf of Mexico points to an even more distant source—the Sahara Desert in Africa. A team of researchers led by John J. Walsh of the University of South Florida promoted this hypothesis in an article titled, "Red Tides in the Gulf of Mexico: Where, When, and Why?" in a 2006 issue of the *Journal of Geophysical Research—Oceans*.

The idea is that winds and storms tend to stir up dense clouds of particles from the dry soil of the Sahara Desert. These dust clouds are thick enough to be viewed with satellite cameras as well as ground-based equipment. Easterly trade winds blow the particles all the way across the Atlantic Ocean, and some of the particles settle in the Gulf of Mexico. A significant component of this dust is iron, a common element in many kinds of soil.

What happens next is a complicated biological interaction. According to the hypothesis, a bacterium known as *Trichodesmium* uses this iron to "fix" nitrogen in the water, making it usable to other marine organisms. (This process is similar to the nitrogen fixation by which bacteria incorporate nitrogen into compounds such as ammonia, providing a critical source of nitrogen for plant life.) Benefiting from this process are opportunistic algae, along with other marine life. Walsh and his colleagues propose that one of the results of this chain of events is the triggering of *Karenia brevis* blooms, even in relatively nutrient-poor water. Such blooms may be common in other areas enriched by iron-containing dust deposits. In their article, the researchers wrote that mysterious *Karenia* blooms "have caused toxic red tides in similar coastal habitats of other western boundary currents off Japan, China, New Zealand, Australia, and South Africa, downstream of the Gobi, Simpson, Great Western, and Kalahari Deserts, in a global response to both desertification and eutrophication."

Marine scientists have yet to determine which of these algal bloom triggers is more important. Perhaps the real trigger is still unknown, and will not be found until researchers probe further into this frontier of science.

MONITORING ALGAL BLOOMS

Even if the exact cause of many algal blooms has not been pinpointed, marine scientists can perform a valuable service by keeping track of algae populations. When dangerous levels of harmful algae appear, scientists warn government officials, who impose fishing and harvesting restrictions so that no contaminated seafood ends up on anybody's dinner plate. Some "red tides" are obvious by the discoloration of water, but since some species are not as easily visible as others—and even relatively mild populations of certain toxic species are worrisome—careful measurements are needed. To monitor algae populations, scientists take samples of coastal waters at a range of sites, and count the number of cells.

In the United States, the Center for Sponsored Coastal Ocean Research fosters projects to monitor algae blooms and, if harmful blooms arise, to respond to the situation. The Center for Sponsored Coastal Ocean Research is one of five centers belonging to NOAA's National Centers for Coastal Ocean Science, which support scientific research and applications toward maintaining and improving the resources along the nation's shorelines. One of the most important programs is called Monitoring and Event Response for Harmful Algal Blooms (MERHAB).

Olympic Region Harmful Algal Blooms (ORHAB) Project

The five-year project, initially scheduled to run from 2000 to 2005, focused on a type of clam known as the razor clam (*Siliqua patula*), which is susceptible to domoic acid, a toxin that comes from *Pseudo-nitzschia,* a genus of algae prevalent in the waters off Washington's Olympic Peninsula. Consumption of contaminated clams can result in a disease called amnesic shellfish poisoning, which attacks the nervous system causing confusion in the victim and, in rare cases, leads to death. Blooms of *Pseudo-nitzschia* are new to Washington—the first recorded *Pseudo-nitzschia* bloom occurred in 1991, but they have continued to strike ever since, similar to the "red tides" of southern New England following the 1972 event.

One of the project's main goals was to develop an adequate sampling operation. Workers collected water once or twice a week from beaches and near the shore at seven locations along the western coast of the Olympic Penin-

To help local authorities and health officials monitor their coastal waters for any signs of harmful algal blooms, MERHAB funds projects to develop and implement monitoring systems. Early studies focused on harmful algal blooms in the Chesapeake Bay, Florida's coastal waters, and around the Olympic Peninsula in Washington. Other projects have bolstered efforts to observe "red tides" in the Gulf of Mexico, toxic freshwater algae in the Great Lakes region, and algae that produce domoic acid, a toxin that has been implicated in a disease called *amnesic shellfish poisoning* (the first medical cases of which occurred during a 1987 outbreak in Canada).

The Olympic Peninsula project began in 2000 and is one of the most impressive successes. This project, called Olympic Region Harmful Algal Bloom (ORHAB), was a partnership of a large number of organiza-

sula. Technicians tested the samples, counting the number of algae and measuring the concentration, if any, of toxin. The collection also sometimes included a few razor clams, in order to determine if the animals' tissues had any signs of domoic acid. New technologies, such as tests that quickly determined algae populations and toxin concentrations, enhanced the speed and accuracy of the monitoring procedures.

Information collected in the sampling process made it easy for officials to act rapidly and decisively when harmful algal blooms occurred. It also prevented any wasteful decisions to restrict harvesting "just in case"—officials knew if a bloom was occurring or not. Because of the money saved by the monitoring operations, the Washington state legislature voted in 2003 to fund the system by imposing a special tax on shellfish licenses (required by those who harvest shellfish in the state). The shellfish and harvesters are doing well, and as reported in the March 16, 2007, ORHAB meeting notes, "Funding from the shellfish license surcharge has exceeded expectations, and there is more money going to the Washington State Department of Health than anticipated."

tions, including the University of Washington, Washington Department of Fish and Wildlife, Pacific Shellfish Institute, and many others, including two Native American tribes—Makah Tribe and Quinault Indian Nation—who live and fish around the Olympic Peninsula. With MERHAB funding, these organizations developed a system for monitoring species of *Pseudo-nitzschia,* some of which produce domoic acid that can accumulate in certain kinds of shellfish. This monitoring system facilitates judgments on when restrictions on shellfish harvesting are needed, preventing poisoning but also reducing economic losses due to unnecessary closures. As described in the sidebar, the success of this monitoring system prompted government officials to adopt it permanently.

Nationwide tracking of diseases associated with harmful algal blooms is the job of the Centers for Disease Control and Prevention

CDC headquarters in
Atlanta, Georgia *(CDC)*

(CDC), the United States government agency responsible for monitoring outbreaks of disease. An important component of this task falls to the Harmful Algal Bloom-related Illness Surveillance System, which collects data on the occurrences of these diseases. Physicians who discover a patient with an illness such as paralytic or amnesic shellfish poisoning, which is normally associated with harmful algal blooms, can report the diagnosis to the surveillance system. Such notifications alert other physicians, as well as government officials, of the possible existence of toxic contamination. By tracing the past meals of the patients, officials can identify the source and destroy the contaminated products.

FORECASTING HARMFUL ALGAL BLOOMS

Detection of a harmful algal bloom in progress is essential to limit the potential adverse impact to human health. But having an advanced warning would be even better, since officials, fishermen, and vacationers could plan ahead and be prepared. Accurate warning will require marine scientists to be able to forecast harmful algal blooms, similar to weather forecasters predicting future storms.

The research discussed in the section "Triggering Runaway Growth" will likely become important in the future development of forecasting systems. If researchers can identify a bloom's trigger, then early detection is possible, perhaps even easy. Studying the fluctuations of nutrients will also be critical.

Another strategy involves obtaining more extensive data. Increasing the sampling frequencies and the number of locations sampled would help, but would be extremely expensive and time-consuming. A more efficient procedure is to gain a more encompassing view with the use of satellite data. The technology to do this already exists, and a satellite-based forecasting system for the Gulf of Mexico has already been implemented. NOAA and the National Aeronautics and Space Administration (NASA) conduct this system, along with help from state and local agencies in the area.

The focus of the Gulf of Mexico Harmful Algal Bloom Operational Forecast System is *Karenia brevis,* the instigator of many "red tides." Satellite images reveal the location and extent of *Karenia* blooms, and whether the bloom is increasing in magnitude or growing smaller. The system also incorporates observations obtained by field workers, which help to confirm satellite observations. After pinpointing the present locations of blooms, scientists analyze prevailing wind and ocean current patterns that may carry *Karenia* to other areas by such and such a time. Forecasters convey these predictions to officials and researchers in the potentially impacted regions, warning them of possible events. For example, forecasters issued many warnings during 2005, especially for Florida.

This satellite forecast system works for *Karenia brevis,* but would not be effective for the detection of most other organisms. The problem is that many species of algae do not color the water as vividly as those responsible for "red tides." Unless researchers can develop instruments to observe these species remotely, satellite forecasts have limited utility.

Satellite observation is not the only possible solution. An alternative strategy that some researchers are pursuing is the development of better models of algae population growth, nutrients, bloom triggers, and the processes of toxin generation and accumulation in shellfish and other marine life. One project that aims to study harmful algal blooms of *Alexandrium fundyense* is the Gulf of Maine Toxicity (GOMTOX) program.

GOMTOX began on September 1, 2006. NOAA's Center for Sponsored Coastal Ocean Research is providing funding for the five-year program. The principal investigator is Donald M. Anderson at WHOI, joined by colleagues at WHOI and several other institutions, including NOAA, the Food and Drug Administration (FDA), and the University of Maine. One of the prime motivators of this project was the red tide of 2005 in the New England region. This harmful algal bloom rivaled the 1972 event in severity.

The Gulf of Maine stretches from Cape Cod, Massachusetts, all the way to the southern tip of Nova Scotia in Canada. Gulf of Maine shores include all the coastline of Maine and New Hampshire, along with the coastline of Massachusetts north of Cape Cod, and parts of the Canadian provinces of Nova Scotia and New Brunswick. This vast gulf is rich in shellfish, worth millions of dollars every year to commercial fisheries. The harmful algal blooms of *Alexandrium fundyense* have put a serious dent in the local economy, as well as continually posing the threat of paralytic shellfish poisoning.

In a press release posted by ScienceDaily and dated October 30, 2006, Anderson noted some of the problems the study will address. "We don't understand the linkages between bloom dynamics and toxicity in waters near shore versus the offshore," he said, "nor do we know how toxicity is delivered to the shellfish in those offshore waters. An additional challenge is the need to expand modeling and forecasting capabilities to include the entire region, and to transition these tools to operational and management use."

As GOMTOX proceeds, researchers are sampling the waters during extensive cruises. They are also using autonomous vehicles and buoys to extend the sampling range. Once enough data has accumulated, project scientists hope to develop mathematical models that will permit officials to forecast the time and magnitude of future blooms. Anderson said, "We will be working closely with federal, state and local officials, resource managers and shellfishermen to synthesize results and disseminate the information and technology. Our ultimate goal is to transition scientific and management tools to the regulatory community for operational use. This project covers the entire Gulf of Maine, including the Bay of Fundy, so there are many affected user groups, communities, and industries who stand to benefit."

NIPPING A BLOOM IN THE BUD

Improved forecasting models will give coastal communities a better chance of preparation. Perhaps by adjusting fishing schedules and limiting the closures and restrictions to an absolute minimum, officials can drastically reduce the impact of harmful algal blooms in the future. But some researchers at the frontier of marine science would like to do better than just forecasting the episodes. If a community knows a harmful algal bloom is imminent or in progress, maybe they can do something to prevent it from evolving into something worse.

Prevention and control offer much more reward than monitoring and forecasting. In his 2004 *Oceanus* article, Solow wrote, "In the U.S. and other developed countries, monitoring efforts and fishery closures have reduced the incidence of human illness caused by toxic algae. However, both monitoring and closures have economic costs that can be substantial. Perhaps the most striking example of this is the complete loss of the wild shellfish resource in Alaska—which once produced 5 million pounds annually—to persistent paralytic shellfish poisoning." Stopping a harmful algal bloom from starting, or controlling one that has, would save a lot of money and resources.

Research on bloom control and prevention is just getting under-way, and some scientists are not optimistic. In 2008, scientists released a report, *Harmful Algal Bloom Research, Development, Demonstration, and Technology Transfer National Workshop Report,* edited by NOAA researcher Quay Dortch and colleagues, based on a meeting of research-ers and officials at Woods Hole, Massachusetts. The report summarized current research on harmful algal blooms and made recommendations for future projects. Many of the projects discussed in this chapter re-ceived favorable mention. But the report described bloom prevention or control as "challenging and controversial." The panel went on to note, "The concept of control refers to strategies that kill HAB [harmful algal bloom] organisms or destroy their toxins directly, remove cells and toxins from the water column physically, or limit the growth and proliferation of the organisms. These strategies intend to reduce the impacts of HABs on people and commerce by targeting the causative agents themselves, reducing their numbers, and minimizing their ef-fects on the environment and commercially important resources. This is one area where HAB science is rudimentary and slow moving."

A major concern of this research is that good intentions do not al-ways lead to the desired results—and sometimes do more harm than good. Unintended consequences are especially prominent in activities that concern mechanisms and processes that scientists do not yet fully understand. The 2008 National Workshop Report described a public meeting in 2006 that discussed harmful algal bloom research in the Gulf of Mexico, in which residents "expressed a range of attitudes toward HAB control. Some were highly supportive of bloom control efforts, while others expressed concern over interfering with natural processes and potential negative impacts on environmental or public health." For example, reducing or limiting nutrients in the water might prevent blooms from developing, but this might also have serious consequences for other marine-life-forms that depend on the same resources. And declines in the population of one species can lead to declines in their predators. A chain reaction might follow, resulting in serious losses of fish and other marine resources.

Although bloom prevention and control methods are not ready for widespread applications, researchers are probing for possible approaches. One of the simplest strategies is to kill the algae by sinking them. Photosynthetic organisms require sunlight that is only available near the surface, so if researchers can find something to send algae to the bottom, they can control a bloom. A possibility for such an "anchor" is clay.

Clay can be effective for this job because it is heavy and sticky. In the late 1980s, Japanese fishermen used clay in a desperate and successful attempt to thwart an algae bloom that threatened fish stocks. After moist clay was spread on the surface, the blooms died off. Korean fish-farmers have also adopted this approach.

But the clay strategy is expensive, and scientists are still not sure of the environmental impacts. According to the 2008 National Workshop Report, "In countries such as Korea, where a fish-farming industry worth hundreds of millions of dollars is threatened by HABs, this control strategy makes sense economically and socially, and so the work has progressed and clay dispersal is now a part of standard HAB management and response. In other geographic areas, the cost/benefit rationale is not as clear. For example, research on clay mitigation has proceeded quite far in the United States, but a significant barrier exists with respect to the ability to obtain permits, environmental clearances, funds, and society's agreement to employ this strategy on more than an experimental scale."

CONCLUSION

No one is certain whether harmful algal blooms are increasing in number because historical data are sketchy. But some regions, such as the Olympic Peninsula in Washington and New England, are certainly experiencing blooms that they have not had to deal with until recently. Considering these and other similar cases, the problem seems to be getting worse.

Scientists have learned a lot about the species of algae responsible for most harmful algal blooms. With this knowledge, officials can monitor coastal waters and ban shellfish harvesting during times when contamination may occur. As monitoring instruments and technology improve, the number and length of closures may be reduced to a minimum.

Much additional research will be necessary before scientists and officials can develop the best possible solution—control and prevention measures. The 2008 National Workshop Report was not optimistic, observing that there are "no examples of marine HAB control being practiced in the United States at the present time and only a few techniques are being used on a small scale in freshwaters. At the current pace of research and development, options for bloom control may not be in place for many years unless a concerted effort is made to encourage and promote these kinds of studies."

One option that has gathered some attention lately is biological control—attacking the algae with other organisms. Unleashing an algae "enemy" or predator at the appropriate moment could potentially eliminate an imminent or incipient bloom. If the attacker is an organism found naturally in the environment, there would seem to be little risk for environmental damage. The only effect would be a short-term, artificial elevation in the organism's population.

Researchers at Scripps Institution of Oceanography recently found a candidate. Xavier Mayali, then a graduate student, and Peter J. S. Franks and Farooq Azam found that *Roseobacter* clade-affiliated (RCA) bacteria, which are common in the oceans, seem to be involved in the natural termination of algal blooms of the dinoflagellate *Lingulodinium polyedrum*. Although the bacteria are about 25–30 times smaller, Mayali and his colleagues conducted laboratory tests that showed the bacteria attaching to the dinoflagellates and destroying them. The researchers published their findings, "Cultivation and Ecosystem Role of a Marine *Roseobacter* Clade-Affiliated Cluster Bacterium," in a 2008 issue of *Applied and Environmental Microbiology.*

These experiments required delicate laboratory manipulations in which Mayali and his coworkers are highly skilled. Although it is not certain that these bacteria perform the same functions in the wild, the researchers sampled algal blooms from all over the world, and found correspondingly large populations of RCA bacteria at these sites, as if the bacteria's numbers were increasing along with their "prey." In a Science-Daily news release posted on May 6, 2008, Franks spoke of the possible applications: "From a practical point of view, if these RCA bacteria really do kill dinoflagellates and potentially other harmful algae that form dense blooms, down the road there may be a possibility of using them to mitigate their harmful effects."

Algal blooms are a part of nature and, as such, humans who live in coastal areas will continue to have to deal with them as in the past. But research on the life cycle, biochemistry, and distribution of algae has helped scientists gain a better understanding of the causes and consequences of harmful algal blooms. And at the frontier of marine science, a few researchers are inching their way toward the ability to suppress these coastal menaces.

CHRONOLOGY

1542	Spanish explorer Álvar Núñez Cabeza de Vaca (ca. 1490–1557) writes a report of travels in America, noting a phenomenon that was probably a harmful algal bloom.
1773	Danish naturalist Otto Friedrich Müller (1730–84) observes dinoflagellates under a microscope.
1793	Captain George Vancouver (1757–98) provides one of the earliest descriptions in America of paralytic shellfish poisoning when some of his crewmen become ill after eating mussels.
1870s	German scientist Friedrich Ritter von Stein (1818–85) publishes excellent dinoflagellate illustrations and descriptions, furthering the classification of these organisms.
1972	Harmful algal bloom associated with paralytic shellfish poisoning extends for the first time into New Hampshire and Massachusetts in a major outbreak. Following the 1972 episode, blooms recur from time to time in this region.

1987	Officials describe a new disease, amnesic shellfish poisoning, when about 100 people in eastern Canada become ill after eating shellfish harvested from around Prince Edward Island. The toxin is domoic acid, which attacks the brain.
1998	Citing millions of dollars in economic losses, the United States government passes the Harmful Algal Bloom and Hypoxia Research and Control Act, calling for monitoring, assessments, and research.
2000	NOAA and NASA scientists begin issuing harmful algal bulletins in the attempt to forecast *Karenia brevis* blooms in the Gulf of Mexico.
2004	United States government extends its 1998 legislation on harmful algal blooms with the Harmful Algal Bloom and Hypoxia Amendments Act.
2005	A major "red tide" strikes the Gulf of Mexico, striking the western coast of Florida particularly hard.
2006	Gulf of Maine Toxicity (GOMTOX) program, a five-year study of harmful algal blooms associated with *Alexandrium fundyense,* begins.
2008	Scripps Institution of Oceanography researchers Xavier Mayali, Peter J. S. Franks, and Farooq Azam publish a report on bacteria that attack dinoflagellates.

FURTHER RESOURCES

Print and Internet

Anderson, Donald M. "The Growing Problem of Harmful Algae." *Oceanus.* November 12, 2004. Available online. URL: http://www.whoi.edu/oceanus/viewArticle.do?id=2483. Accessed June 9, 2009. Anderson discusses the threats of harmful algae blooms.

Center for Sponsored Coastal Ocean Research. "Harmful Algal Blooms." Available online. URL: http://www.cop.noaa.gov/stressors/extremeevents/hab/. Accessed June 9, 2009. The Center for Sponsored Coastal Ocean Research, one of NOAA's National Centers for Coastal Ocean Science, presents the results of their studies of harmful algal blooms, including the environmental and economic impacts.

Dortch, Q., D. M. Anderson, D. L. Ayres, and P. M. Gilbert. *Harmful Algal Bloom Research, Development, Demonstration, and Technology Transfer National Workshop Report.* Available online. URL: http://www.whoi.edu/fileserver.do?id=43464&pt=10&p&19132. Accessed June 9, 2009. Based on a meeting of researchers and officials at Woods Hole, Massachusetts, the report summarizes current research on harmful algal blooms and makes recommendations for future projects.

Fillmore, Randolph. "Red Alert Red Tide." *USF Magazine.* Winter, 2005. Available online. URL: http://www.sciencescribe.net/articles/Red_Alert_redtide.pdf. Accessed June 9, 2009. This article offers a vivid description of the terrible red tide experienced on Florida's Gulf coast in 2005.

Lax, Alistair J. *Toxin: The Cunning of Bacterial Poisons.* Oxford: Oxford University Press, 2005. Although the book is not about marine toxins, it contains a great deal of relevant information about how toxic substances have been discovered and how they exert their harmful effects.

Lee, Robert Edward. *Phycology.* 4th ed. Cambridge: Cambridge University Press, 2008. Phycology is the study of algae. This book offers an excellent introduction for advanced students to this branch of biology.

Mayali, Xavier, Peter J. S. Franks, and Farooq Azam. "Cultivation and Ecosystem Role of a Marine *Roseobacter* Clade-Affiliated Cluster Bacterium." *Applied and Environmental Microbiology* 74 (2008): 2,595–2,603. The researchers have found that *Roseobacter* clade-affiliated (RCA) bacteria seem to be involved in the natural termination of algal blooms of certain dinoflagellates.

Meinesz, Alexandre. *Killer Algae: The True Tale of a Biological Invasion.* Chicago: University of Chicago Press, 1999. The spread of certain algae can have dire effects not only from harmful algal blooms but also from ecological damage. This book documents serious ecosystem changes that resulted from the accidental introduction in the 1980s of the alga *Caulerpa taxifolia* into the Mediterranean Sea.

National Oceanic and Atmospheric Administration. "Harmful Algal Bloom Operational Forecast System." Available online. URL: http://tidesandcurrents.noaa.gov/hab/. Accessed June 9, 2009. Information on the forecasting techniques and links to past harmful algal bloom bulletins can be found here.

Pittman, Craig. "Red Tide Blame Points to Mississippi." *St. Petersburg Times.* November 8, 2007. Available online. URL: http://www.sptimes.com/2007/11/08/State/Red_Tide_blame_points.shtml. Accessed June 9, 2009. The article discusses the possibility that the Mississippi River may be contributing to Florida's red tides.

Rapport, Josh. "Dinoflagellate Research Paper." Available online. URL: http://www.mbari.org/staff/conn/botany/dinos/paper.htm. Accessed June 9, 2009. This interesting article on dinoflagellates briefly discusses how the organisms were discovered, how they are classified, their biology, and the various toxins they produce.

ScienceDaily. "Red Tide Killer Identified: Bacteria Gang Up on Algae, Quashing Red Tide Blooms." News release, May 6, 2008. Available online. URL: http://www.sciencedaily.com/releases/2008/05/080501 125429.htm. Accessed June 9, 2009. Scientists discover organisms that may terminate harmful algal blooms.

———. "Red Tide Models and Forecasts To Be Expanded in Gulf of Maine." News release, October 30, 2006. Available online. URL: http://www.sciencedaily.com/releases/2006/10/061017084608.htm. Accessed June 9, 2009. WHOI and other agencies announce a plan to study algal blooms and toxins in the Gulf of Maine.

Solow, Andrew R. "Red Tides and Dead Zones." *Oceanus.* December 22, 2004. Available online. URL: http://www.whoi.edu/oceanus/viewArticle.do?id=2487. Accessed June 9, 2009. Solow discusses eutrophication and dead zones created by algal blooms.

Stumpf, Richard P., R. Wayne Litaker, Lyon Lanerolle, and Patricia A. Tester. "Hydrodynamic Accumulation of *Karenia* off the West Coast of Florida." *Continental Shelf Research* 28 (2008): 189–213. The researchers promote the idea that the huge outflow of the Mississippi River may be contributing to algal blooms associated with *Karenia brevis* and related species all the way to Florida.

Walsh, J. J., J. K. Jolliff, B. P. Darrow, J. M. Lenes, S. P. Milroy, A. Remsen, et al. "Red Tides in the Gulf of Mexico: Where, When, and Why?" *Journal of Geophysical Research—Oceans* 111 (2006). Available online. URL: http://www.agu.org/pubs/crossref/2006/2004JC002813. shtml. Accessed June 9, 2009. The researchers discuss the causes of red tides in the Gulf of Mexico.

Woods Hole Oceanographic Institution. "GOMTOX: Dynamics of *Alexandrium fundyense* Distributions in the Gulf of Maine." Available online. URL: http://www.whoi.edu/gomtox/. Accessed June 9, 2009. Details of the Gulf of Maine Toxicity (GOMTOX) program are available on this page, including the details of the proposal, progression time line, news, and a list of participants.

Yohe, Evelyne. "Hunting Dangerous Algae from Space." Available online. URL: http://earthobservatory.nasa.gov/Study/Redtide/. Accessed June 9, 2009. Archived at NASA's Earth Observatory Web site, this article, dated July 9, 2002, offers a concise and interesting introduction to the use of satellites in monitoring harmful algae blooms.

Web Sites

Bigelow Laboratory for Ocean Sciences: Toxic and Harmful Algal Blooms. Available online. URL: http://www.bigelow.org/hab/. Accessed June 9, 2009. This Web site is full of information on harmful algal blooms, including the species involved and where they are located, the diseases they cause, and how the toxins cause disease.

Centers for Disease Control and Prevention: Harmful Algal Blooms. Available online. URL: http://www.cdc.gov/hab/. Accessed June 9, 2009. Charged with collecting information and investigating diseases associated with harmful algae, CDC maintains these pages describing their data collection, research, and public health alerts.

Woods Hole Oceanographic Institution: Harmful Algae. Available online. URL: http://www.whoi.edu/redtide/. Accessed June 9, 2009. This Web site explains what harmful algae blooms are and what they do, and provides information on the species involved, where they are found, and what researchers are doing to investigate the problem.

FINAL THOUGHTS

The continued efforts of oceanographers will undoubtedly lead to further advances in the subjects covered in this book, as well as many other important questions in marine science. But there is a critical variable that will affect many of these topics in as yet unknown ways. This variable is global climate change—and it is a serious concern to everyone.

A scientific organization known as the Intergovernmental Panel on Climate Change (IPCC) is one of the primary groups that monitor the situation. In a 2007 report, *Climate Change 2007,* the panel said that over the 100-year period between 1906 and 2005, global surface temperature increased by 1.33°F (0.74°C). This is only an average—some areas, such as parts of the southeastern United States, have cooled, whereas places such as Alaska have experienced a higher than average rate.

What is causing global warming? Recent research has focused on the increase of "greenhouse" gases such as carbon dioxide, issuing from human activities such as combustion of fossil fuels. Such gases in the atmosphere absorb infrared radiation, which is a low-frequency radiation emitted by warm bodies as they cool off. The Sun's rays warm the Earth's surface, causing it to emit infrared radiation, but an increase in atmospheric carbon dioxide may be trapping this radiation, resulting in a rise in temperature. This process is similar to the warming effect of a greenhouse, in which panes of glass let in high-frequency radiation from the Sun but block infrared emission.

No one is certain how much human activity is contributing to global warming. Temperatures have fluctuated over much of Earth's recent history, resulting in periods of cold weather and glacier buildup known as ice ages, and warming trends, such as occurred during the end of the last major glacial period, about 10,000–15,000 years ago. These warming trends long preceded the 18th-century Industrial Revolution and fossil fuel combustion, so they obviously had another cause. Yet the present increases in "greenhouse" gases can clearly have an impact, and the recent temperature rise has been rapid and unaccompanied by any significant event other than increases in industrial emissions. In their 2007 report, IPCC says that they have a "very high confidence that the global average net effect of human activities since 1750 has been one of warming." By "very high confidence," IPCC means "at least a 9 out of 10 chance of being correct."

One result of warming trends may be a significant rise in sea level. The melting of glaciers and other land ice releases a considerable amount of water, most of which falls into the ocean. Rising sea levels could alter coastlines, inundating low-lying areas.

Measurements of researchers at the National Oceanic and Atmospheric Administration (NOAA) and elsewhere indicate a rise in the global mean sea level at a rate of 0.07 inches (0.17 cm) per year over the last century. This is not a particularly frightening rate, and most of this rise is due to thermal expansion, a natural process by which the volume of substances, including water, increases with rising temperature. As the ocean gets warmer, its volume increases.

But decreases in glaciers and ice in the sea is also occurring. IPCC's 2007 report refers to satellite observations that indicate a reduction of sea ice in the Arctic Ocean at an annual rate of 2.7 percent. The United Nations Environment Programme announced on March 16, 2008, that data from 30 glaciers in nine mountain ranges show a thinning since 1980 about the equivalent of 34.4 feet (10.5 m) of water. If all the land ice in the world melted, sea levels would rise about 213 feet (65 m). (Melting sea ice does not increase the sea level since floating ice displaces its equivalent weight in water.)

No one is suggesting that all land ice will soon disappear, which would cause catastrophic flooding. But a rise in sea level due to the continual melting of glaciers can be expected. In 2007, IPCC estimated a rise in sea level in the range of 0.59 feet (0.18 m) to 1.97 feet (0.6 m) by the year 2100.

This estimate was designed to be conservative. A more accurate assessment must take into account a number of different factors, including future temperature changes, the rate of land ice melting, and evaporation and atmosphere vapor content. Because of the complexity and global scale of these calculations, combined with the uncertainty of the future, researchers are struggling to improve their models.

In a 2008 issue of *Science,* University of Colorado scientist W. Tad Pfeffer and his colleagues published their estimate of future sea levels. The paper, "Kinematic Constraints on Glacier Contributions to 21st-Century Sea-Level Rise," focused on Greenland and its glaciers because the variables here are more easily observed and "well constrained"— meaning there is not as much uncertainty. The researchers then projected the results of their calculations, and found that the highest plausible sea level rise by 2100 would be 6.56 feet (2.0 m). A more likely scenario is a rise of 2.6 feet (0.8 m).

But even modest increases in sea level would threaten flat coastal regions and islands. The Netherlands, for instance, has a generally low elevation and would face severe flooding. Kiribati, a republic composed of a group of islands in the Pacific Ocean, barely rises above 6.5 feet (1.98 m) in most places, and would be submerged! Low-lying coastal regions of the United States are also threatened, and some officials are already taking action. On November 14, 2008, for instance, California governor Arnold Schwarzenegger requested a study of the potential impact of rising sea levels on the 800-mile (1,280-km) coastline of the state. Schwarzenegger warned, "The longer that California delays planning and adapting to sea level rise the more expensive and difficult adaptation will be."

What kind of adaptation will be required depends on how the world's ocean responds to global climate change. Researchers are beginning to make progress on developing more precise models, but much uncertainty—and fear—remains. A better understanding of ocean and weather systems will alleviate much of the fear and uncertainty and help determine an effective course of action having the least disruption to society. Achieving this goal is one of the most important challenges at the frontiers of marine science.

GLOSSARY

abyssal plain   a relatively flat region of the deep ocean floor

algae   a mostly aquatic group of organisms capable of photosynthesis

amnesic shellfish poisoning   disease caused by consumption of marine organisms contaminated with certain toxins, which attacks the victim's nervous system, causing confusion and loss of memory

archaea   one of the three domains of life, consisting of microscopic, bacteria-like organisms

bathymetry   measurement of water depth

benthic   pertaining to the ocean floor

bioluminescence   light-emitting biochemical process

chemosynthesis   process by which organisms use the energy in chemical bonds to manufacture life-sustaining molecules

continental shelf   the shallow, gently sloping oceanic region extending from the shore of a continent or island up to the point where the depth begins to drop more sharply

cyst   an algal organism in a dormant stage of its life cycle

dead zone   region of the sea with insufficient oxygen to support most marine organisms

deoxyribonucleic acid   a molecule that contains hereditary information in the sequence of its components

dinoflagellates microscopic plankton responsible for many harmful algal blooms

DNA *See* **deoxyribonucleic acid**

El Niño/Southern Oscillation correlated changes in air pressure, winds, and ocean temperatures in the Pacific Ocean that are associated with weather patterns all over the world

ENSO *See* **El Niño/Southern Oscillation**

eutrophication enriched with nutrients

extremophiles organisms that thrive in environments with extremely harsh conditions such as high temperatures or acidity

fathoms units of depth equal to 6 feet (1.83 feet)

fault a fracture in rocks in which displacement has occurred, or in other words, one side has moved relative to the other side

HAB *See* **harmful algal blooms**

harmful algal blooms sudden explosions in certain algae populations that produce toxins and/or cause hypoxia

hydrothermal vents ocean floor outlets of hot water, which the underlying magma has heated

hyperbaric greater than normal pressure

hypoxia condition in which a sufficient supply of oxygen is not available

igneous rocks solidified from magma or lava

magma molten rock

magnetic field region of space in which certain forces attract iron and similar materials and influence the motion of electrically charged particles

magnetometers instruments used to measure the strength or direction of a magnetic field

metabolism   life-sustaining chemical reactions in organisms

meteorology   the study of the atmosphere and weather

mid-ocean ridge   volcanic formation that is created by tectonic plate separation, and extends along the seabed through the oceans

NASA   *See* **National Aeronautics and Space Administration**

National Aeronautics and Space Administration   United States government agency devoted to the development of flight technology and space exploration

National Oceanic and Atmospheric Administration   United States government agency that monitors and studies Earth's oceans, atmosphere, and weather

NOAA   *See* **National Oceanic and Atmospheric Administration**

paralytic shellfish poisoning   disease caused by consumption of marine organisms contaminated with certain toxins, which attacks the victim's nervous system and causes nausea, tingling sensations, and in extreme cases leads to muscle paralysis and death

photosynthesis   the process by which plants use the energy of sunlight to make sugar (carbohydrates)

phytoplankton   algal plankton

plankton   small organisms that tend to drift with the ocean currents

radioactive   exhibiting the property of certain atoms to emit high-energy radiation

rift valley   deep fracture in the mid-ocean ridge, where the tectonic plates are separating

salinity   a measure of the amount of dissolved material such as salts in the water

seamounts   underwater mountains that do not rise high enough to break the water's surface

sediment particles that accumulate on the bottom of an ocean or lake

seismic waves vibrations that occur during earthquakes or sudden movements of large volumes of rock

seismometers instruments used to measure seismic waves

sonar sound navigation and ranging, a method of using sound to map undersea terrain or locating objects in the water

submersible an underwater vessel, usually one with limited mobility that is transported or lowered by a ship

tectonic plates a dozen or so massive slabs of rock, and a few dozen smaller slabs, which cover Earth's surface, extend to a depth of about 62 miles (100 km), and slowly drift in the range of 1–6 inches (2.5–15 cm) per year

teleconnection linkage or correlation of distant weather patterns

thermocline layer of water, lying between the warmer surface layer and the colder water below, in which there is a rapid temperature transition

toxins poisonous substances

turbidity cloudiness in a fluid caused by the presence of particulate matter

wavelength distance between crests (or, equivalently, the distance between any other phase, such as troughs)

FURTHER RESOURCES

Print and Internet

Ballard, Robert D. *The Eternal Darkness: A Personal History of Deep-Sea Exploration.* Princeton, N.J.: Princeton University Press, 2000. Ballard has led many important expeditions, including those that found the RMS *Titanic* and made the initial discovery of hydrothermal vents. In this book, he describes some of the early history of deep-sea diving and then recounts his pioneering adventures.

Ballesta, Laurent, and Pierre Descamp. *Planet Ocean: Voyage to the Heart of the Marine Realm.* Washington, D.C.: National Geographic Society, 2007. Rich in stunning images, each chapter of this book describes a different marine environment, such as the polar seas, coral reefs, and many others.

Earle, Sylvia. *National Geographic Atlas of the Ocean: The Deep Frontier.* Washington, D.C.: National Geographic Society, 2001. Full of maps, photographs, charts, and diagrams, this book reveals the structure and inhabitants of the world's oceans.

Field, John G., Gotthilf Hempel, and Colin P. Summerhayes, eds. *Oceans 2020: Science, Trends, and the Challenges of Sustainability.* Washington, D.C.: Island Press, 2002. This forward-looking book identifies and discusses issues in marine science and ocean conservation that will likely be a priority in the early years of the 21st century.

Kunzig, Robert. *Mapping the Deep: The Extraordinary Story of Ocean Science.* New York: W. W. Norton & Company, 2000. Kunzig relates all the major new discoveries of marine science, including seafloor spreading, undersea mapping, and the many varieties of deep-sea organisms.

Lewis, Jon E., ed. *The Mammoth Book of the Deep: Over 30 True Stories of Danger and Adventure Under the Sea.* New York: Carroll & Graf, 2007. This collection of stories includes Robert D. Ballard's "The Discovery of the *Titanic*," Jacques-Yves Cousteau's "Menfish," and William Beebe's "The Kingdom of the Helmet."

National Oceanic and Atmospheric Administration. "Ocean Explorer." Available online. URL: http://oceanexplorer.noaa.gov/. Accessed June 9, 2009. This Web resource highlights the fascinating expeditions of NOAA's science and education teams. Plenty of videos, photos, and maps are included.

Pearce, Fred. *With Speed and Violence: Why Scientists Fear Tipping Points in Climate Change.* Boston: Beacon Press, 2007. Tipping points are unstable, liable to topple one way or another at any moment. Science writer Fred Pearce describes numerous situations, some of which involve marine science, which threaten to upset critical environmental properties quickly and with long-lasting—and possibly dire—effects.

Prager, Ellen. *Chasing Science at Sea: Racing Hurricanes, Stalking Sharks, and Living Undersea with Ocean Experts.* Chicago: University of Chicago Press, 2008. Oceanography is an exciting field of science, but like all such endeavors, it is not always a bed of roses. Prager describes the thrills as well as the challenges of being a marine scientist.

Roberts, Callum. *The Unnatural History of the Sea.* Washington, D.C.: Island Press, 2007. Roberts chronicles the history of the fishing industry, chronicling the time of abundance in the early days up to the perilously waning fish stocks of today.

Ulanski, Stan. *The Gulf Stream: Tiny Plankton, Giant Bluefin, and the Amazing Story of the Powerful River in the Atlantic.* Chapel Hill, N.C.: University of North Carolina Press, 2008. The Gulf Stream is a strong ocean current flowing between the Gulf of Mexico and the European coast. This book highlights its numerous denizens, along with the many effects it has on marine and coastal environments.

Woods Hole Oceanographic Institution. *Oceanus: The Magazine that Explores the Oceans in Depth.* Available online. URL: http://www.whoi.edu/oceanus/. Accessed June 9, 2009. The online version of *Oceanus* provides research news, feature articles, interviews and quotes with researchers and students, and information on current WHOI initiatives.

Web Sites

Discover Education: Planet Ocean. Available online. URL: http://school.discoveryeducation.com/schooladventures/planetocean/. Accessed

June 9, 2009. Aimed at students, this presentation features pages explaining the ocean and many of its life-forms, including the blue whale, barracuda, and tubeworm.

Exploratorium. Available online. URL: http://www.exploratorium.edu/. Accessed June 9, 2009. The Exploratorium, a museum of science, art and human perception in San Francisco, has a fantastic Web site full of virtual exhibits, articles, and animations, including much of interest to oceanographers and oceanographers-to-be.

How Stuff Works. Available online. URL: http://www.howstuffworks.com/. Accessed June 9, 2009. This Web site hosts a huge number of articles on all aspects of technology and science, including marine science.

National Oceanic and Atmospheric Administration. Available online. URL: http://www.noaa.gov/. Accessed June 9, 2009. NOAA's Web site provides a huge amount of information on weather and climate issues as well as marine science and oceanography. Topics include basic information, observational data, and NOAA research.

ScienceDaily. Available online. URL: http://www.sciencedaily.com/. Accessed June 9, 2009. An excellent source for the latest research news, ScienceDaily posts hundreds of articles on all aspects of science. The articles are usually taken from press releases issued by the researcher's institution or by the journal that published the research. Main categories include Plants & Animals, Earth & Climate, Matter & Energy, and others.

Scripps Institution of Oceanography. Available online. URL: http://www.sio.ucsd.edu/. Accessed June 9, 2009. Scripps's Web site is full of news and research from one of the most active marine science organizations in the world.

Smithsonian Institution: Ocean Planet. Available online. URL: http://seawifs.gsfc.nasa.gov/ocean_planet.html. Accessed June 9, 2009. Ocean Planet was a traveling exhibition of the Smithsonian Institution. This Web site contains the text and most of the panels and images of the exhibition. Topics include Oceans in Peril, Sea People, Ocean Science, and much more.

Woods Hole Oceanographic Institution. Available online. URL: http://www.whoi.edu/. Accessed June 9, 2009. With pages devoted to WHOI's history, research, ships and technology, scientists, education, and news, WHOI's Web site offers a wealth of information about this active oceanographic institution.

INDEX